CW00402816

BIG BEN:
THE GREAT CLOCK AND
THE BELLS AT THE PALACE
OF WESTMINSTER

Frontispiece: Edmund Beckett Denison, designer of the Great Clock of Westminster, the bell Big Ben, and the quarter bells. This photo was used to produce the coloured pastel now in the British Horological Institute. See Plate 14 in the colour plate section. Photo courtesy of Smith of Derby.

BIG BEN:
THE GREAT CLOCK AND THE BELLS AT THE PALACE OF WESTMINSTER

Chris McKay

OXFORD

UNIVERSITY PRESS

OXFORD

UNIVERSITY PRESS

Great Clarendon Street, Oxford OX2 6DP

Oxford University Press is a department of the University of Oxford.
It furthers the University's objective of excellence in research, scholarship,
and education by publishing worldwide in

Oxford   New York

Auckland Cape Town Dar es Salaam Hong Kong Karachi
Kuala Lumpur Madrid Melbourne Mexico City Nairobi
New Delhi Shanghai Taipei Toronto

With offices in

Argentina Austria Brazil Chile Czech Republic France Greece
Guatemala Hungary Italy Japan Poland Portugal Singapore
South Korea Switzerland Thailand Turkey Ukraine Vietnam

Oxford is a registered trade mark of Oxford University Press
in the UK and in certain other countries

Published in the United States
by Oxford University Press Inc., New York

© C.G. McKay 2010

The moral rights of the author have been asserted
Database right Oxford University Press (maker)

First published 2010
Reprinted 2010

All rights reserved. No part of this publication may be reproduced,
stored in a retrieval system, or transmitted, in any form or by any means,
without the prior permission in writing of Oxford University Press,
or as expressly permitted by law, or under terms agreed with the appropriate
reprographics rights organization. Enquiries concerning reproduction
outside the scope of the above should be sent to the Rights Department,
Oxford University Press, at the address above

You must not circulate this book in any other binding or cover
and you must impose the same condition on any acquirer

British Library Cataloguing in Publication Data

Data available

Library of Congress Cataloging in Publication Data

Data available

Typeset by SPI Publisher Services, Pondicherry, India

ISBN 978-0-19-958569-4

3 5 7 9 10 8 6 4 2

FOREWORD

Mike McCann
The Keeper of the Great Clock

My daily walk over Westminster Bridge, threading my way between the throngs of camera-laden tourists, provides me with a daily reminder of just how fascinated children and adults alike are with Big Ben and its Great Clock. Even after 150 years, children seem to love the idea of a giant bell ringing far above them, while adults, in a world of quartz and atomic clocks, are amazed at the accuracy of the 5 ton mechanical clock and how old pennies are used to ensure that it remains so. Personally, despite more than a decade of hurrying to catch my train like every commuter, I still find time to pause and admire the genius of its Architect, Charles Barry, in creating such an iconic tower with its wonderful gold spire and finials and its stunning clock face.

Our Victorian ancestors were equally fascinated and flocked to London in their thousands by steam train to see the construction of the Clock tower, which appeared to rise by magic, being built from the inside with the use of a steam-driven platform and, as a result, virtually no external scaffolding. This was the heyday of Victorian Britain, when architects, designers, and engineers were the celebrities of their day and, just like today, the public were entertained by the highly public quarrels played out in the press between the buildings architect and the designer of Big Ben and the Clock, Edmund Dennison, QC. This culminated in the disastrous failure of the first Big Ben and indeed the subsequent cracking of its replacement. Back in 1844 though, the Chief Lord of Woods and Forest, Sir Benjamin Hall, had promised, '*A noble clock, indeed a king of clocks, the biggest the world had ever seen, within sight and sound of the throbbing heart of London,*' and despite these quarrels and numerous setbacks, Big Ben and its clock triumphed to become one of the worlds most recognized, reliable, and celebrated icons.

Celebrations for its 150th anniversary have ranged from Post Office special edition covers and Royal Mint £5 coins to Chris McKay's book, the first produced by an actual horologist. Readers will find it full of fascinating details and drawings.

Mike McCann, MSc, BEng (Hons), CEng, MInstE. The Keeper of the Great Clock. This figure is reproduced in colour in the colour plate section.

ACKNOWLEDGEMENTS

The following persons are gratefully thanked for their help and contributions. Some contributions were small, some significant, but without such generous help and support, the book would only have been a shadow of this, final, version.

PALACE STAFF

I would particularly like to thank Mike McCann, the Keeper of the Great Clock, who was kind enough to accommodate my visits to the Clock Tower whilst he carried on his busy tasks as Palace Maintenance manager.

Additional thanks go to the Palace clock–makers, who were very obliging in showing me the clock and bells:

Paul Roberson,
Huw Smith,
Ian Westworth.

The help of the following Palace staff is also acknowledged:

Dr. Mark Collins,
Edward Garnier, MP,
Amanda Leck,
Gemma Webb.

The help and assistance of the following indi–viduals and organizations is hereby gratefully acknowledged:

PEOPLE

John Ablott,
Vic Adams,
Don Amphlett,
Robert Ball,
Doug Bateman,
Jonathan Betts,
Mike Bundock,
Graham Chinner,
Keith Cotton,
Paul Craddock,
Mel Doran,
John Edge,
John Fisher,
Stella Haward,
Bill Hibberts,
Richard Johnston,
Les Kirk,
Rodney Law,
Melvyn Lee,
Dickon Love,
Malcolm Loveday,
Michael Maltin,
Tony and Katie Marshall,
Alan Midleton,
Cathy Moss,
Chris Pickford,
John Plaister,
Nigel Platt,

Michael Potts,
Tim Redmond,
Lindsay Schusman,
Mick and Eleanor Skadorwa,
Nick Smith,
Lord Somerleyton,
Mike and Frances Trickett,
John Vernon,
David Vulliamy,
John Warner,
Peter Watkinson,
Ken Weatherhogg,
John Wilding,
Keith Scobie-Youngs.

ORGANIZATIONS AND COMPANIES

Antiquarian Horological Society,
British Horological Institute,
British Library,
Cumbria Clock Company,
East Riding Record Office,
Guildhall Library, Manuscript, Map, and Print Rooms,
Hartlepool Museum,
Institution of Civil Engineers,

London Metropolitan Archives,
National Archives,
National Physical Laboratory,
Parliamentary Archives,
Royal Astronomical Society Library,
Science Museum Library,
Smith of Derby,
Whitechapel Bell Foundry.

I would like to offer my deep thanks to those who read through my script, corrected various errors, and provided valuable inputs from their specialist fields of knowledge:

John Edge,
Malcolm Loveday,
Chris Pickford,
Keith Scobie-Youngs,
Mike Trickett.

PHOTO CREDITS

All modern pictures taken within the Houses of Parliament are courtesy of the Palace of Westminster.

Pictures and figures accredited to the NPL are © Queen's Printer and Controller of HMSO, 1976.

CONTENTS

INTRODUCTION

My interest in clocks began early. I was always taking things apart to see how they worked, and there seemed an abundance of alarm clocks and the like to tinker with. When I was 11 years old, I often stared up at a derelict clock tower that overshadowed the school's playing field, wondering if I could get the clock working again. This became a reality when I was 19, and led to a lifelong interest in the practical, technical, and historical aspects of turret clocks.

When I was 13, the family moved to Greenford in West London and I would cycle around, exploring. One day I discovered a small library hidden in a residential area of Southall; here I found a book *A Rudimentary Treatise on Clocks, Watches, and Bells* by a Lord Grimthorpe. The 1903 copy was battered and falling apart, but I was fascinated by its contents, explanations of how church clocks worked, and a full description and history of the Great Clock at Westminster.

Throughout a career in the electronics industry, followed by a second career in teaching, turret (tower) clocks were always my main interest. A third, part-time, career followed, working on turret clocks and delivering related training. Since university, I had been involved in church bell ringing, a useful pursuit that brought me in contact with the practical side of bell maintenance and offered the chance to see many turret clocks. I was always involved in some way with clock restorations, lectures, writing articles, advising, and historical research. One character

into whom I researched extensively was this same Lord Grimthorpe, Edmund Beckett Denison, the designer of the Great Clock at Westminster. Amassing and interpreting information led to the idea of a book on Denison. It was then that I realized that although there were plenty of booklets, chapters in books and references to Big Ben, there was, however, no definitive work that drew all the information together. The task was then obvious: produce the book. I began with just the clock movement in mind but my subject soon expanded to include the bells, the tower and, most importantly, the people involved. Apart from Vaudrey Mercer's book on E.J. Dent, previous published studies in the main lacked technical detail and originality, and repeated errors from previous works. In this book I hope to offer a wholly new account, thoroughly researched from original sources.

I have tried to make the contents as accessible to as wide a readership as possible. I expect that this book will be of most interest for the practical and antiquarian horologist, along, of course, with those interested in bells.

Since the Great Clock and everything associated with it was designed using old Imperial units, I have decided to retain these in the text. However, appendixes provide data listed in both Imperial and metric units, and provide conversion information.

I have integrated the references to source material into my text where possible, to avoid extensive footnotes. Details are also given at the end of the book. A lot of information was taken from two sets of letters on the clock and bells in the National Archives. Although much of this was published in various Parliamentary Papers, the order and content was not always easy to follow.

The Illustrated London News, The Times newspaper and Hansard also provided much information, along with Edmund Beckett Denison's books.

I hope that the reader enjoys my history of the Great Clock and the bells as much as I did while researching and compiling the book.

GLOSSARY

Arbor	Horological term for axle
Armchair	Incorrect term for double-framed
Automatic winder	Electrically powered device to wind clock
Barrel	Wood or metal cylinder, around which a weight line is wound
Bell crank	Lever to transfer the pull of a wire through 90°
Bell hammer	Hammer to sound bell
Bevel gear	Set of gears to transfer turning of a rod through 90°
Birdcage	Type of clock frame
Bob	Weight on the end of a pendulum
Bushing	Brass bearing in which pivots run
Carillon	Term casually used to describe a tune barrel. Strictly, a set of 23 or more bells
Cast iron	Iron cast in a mould
Centre wheel	Wheel in going train, turning once an hour
Click	Device to stop a wheel turning backwards
Compensation pendulum	Pendulum specially constructed to keep correct time at different temperatures.
Count wheel	Wheel to set the number of blows struck when the clock strikes
Deadbeat	Type of escapement
Dial	The proper name for what some people call the face
Differential (epicyclic)	Gears used in a maintaining power
Dog clutch	Device to engage or disengage a drive
Double-framed	Type of turret-clock frame
Double three-legged gravity	Type of escapement
End-to-end	Early type of turret clock frame
Epicyclic	Sun and planet gears used in a maintaining power and automatic winders
Escape	Shortened form for escapement
Escapement wheel	Wheel on which the escapement acts
Escapement	Device to release one tooth at a time and to impulse pendulum
Flatbed	Type of turret-clock frame
Fly	Fan-shaped device to limit the speed of striking

Foliot	Precursor to the pendulum: a weighted arm first pushed one way then another
Four poster	Type of clock frame
Frame	The means to contain the wheels in a clock, usually of iron, wood, or brass
Friction clutch	Device to set hands to time
Great wheel	The largest wheel in a clock train
Horology	The craft and science of making clocks and watches
Huygens endless chain	An endless chain used in an automatic winder
Impulse	The action of an escapement, giving a push to a pendulum
Jack	Decorative figure or automaton who strikes a bell
Leading-off rod	Rod connecting the clock to a dial
Leading-off work	Collection of rods and gears to connect the clock to a dial
Line	Line or rope from which the driving weight hangs, usually of galvanized steel
Maintaining power	Device to keep clock running whilst it is being wound
Monkey-up-a-rope	Automatic winder where the motor climbs up a chain
Motion work	Reduction gears behind a dial to drive the hour hand from the minute hand
Override switch	Safety switch, which disables an automatic winder in the event of a switch failure
Pallets	Parts of the escapement which impinge on the escape wheel teeth
Pendulum	The device that keeps the time
Pinion	Small gear of 12 teeth or fewer
Pinwheel	Type of escapement
Pivot	Part about which a wheel turns
Plate and spacer	Type of clock frame
Pulley	Used to guide weight lines or suspend the driving weight
Quarter chiming	A more informal term for quarter striking
Quarter striking	Chimes that sound every 15 minutes
Rack	Device to count number of blows to be sounded at the hour
Ratchet	Gear wheel with saw-like teeth, arrested by a click
Recoil	Type of escapement
Regulating nut	Nut to adjust the timekeeping of a pendulum
Setting dial	Internal dial to enable external dial to be set to time
Snail	Cam to set how far a rack falls
Striking	The sounding of a bell at the hour
Suspension spring	Thin spring from which the pendulum hangs
Synchronous motor	Electrical motor driven by the mains
Ting tang	Quarter chime sounded on two bells of different notes

Train	Collection of gear wheels
Train bar	Bar to retain a train of gears
Tune barrel	Device to play tunes on bells
Turret clock	Clock with a public dial or sounding on a bell
Up	Bells supported mouth up and ready to ring
Verge	Type of escapement
Waiting train	Type of electrical turret clock
Warning	The release of the strike or a chime train a few minutes before striking or chiming
Wheel	Gear with 13 or more teeth
Winding jack	Reduction gear to make winding a clock easier
Wrought iron	Iron forged by a black smith

LIST OF PLATES

CHAPTER ONE
'BIG BEN'—A BRIEF HISTORY

'Big Ben' is the icon of Great Britain in the same way that the Eiffel tower and the Statue of Liberty identify France and the United States. It is the hour bell that is called Big Ben: the clock is properly called The Great Clock of the Palace of Westminster, while the tower is known as the Clock Tower. Sometimes people call the tower St Stephen's Tower or the Albert Tower, but both names are incorrect. However, in popular parlance the term 'Big Ben' includes tower, bells, and clock.

Westminster first became a place of habitation when the Abbey was founded, around 960 AD. Edward the Confessor set up court there about 1040, and the Abbey formed part of his Royal Palace. Apart from the Abbey, the most famous building is Westminster Hall, which was built by William II, known as William Rufus. Along with the Royal Palace came other functions of a seat of government: an Exchequer, a Treasury, and Courts of Justice. Eventually, in 1265, nobles were summoned to consider important affairs of state and thus the first Parliament was formed.

The first clock at Westminster was installed in a clock tower around 1290. This was built close to the Palace at the north end of the Westminster Hall, and struck the hour on a large bell. The tower was demolished in 1707 and the bell sold to St Paul's cathedral.

Eventually, the monarchy moved from Westminster to Whitehall, but the important functions of government remained, including a regular meeting of nobles and commoners; the Parliament. In time, the buildings that made Westminster Palace became know as the Houses of Parliament.

In 1834, the Palace was a collection of buildings from many different eras; developed and

FIG I.I 'Big Ben': the most famous clock tower in the world.

I

added to piecemeal, as needs directed. In October that year a fire started that destroyed most of the Palace, but Westminster Hall was saved. It was decided that the Palace should be pulled down and a new building put up. The design was to be in the Gothic or the Elizabethan style and was put out to public competition. Charles Barry's design won and, as part of his prize, Barry was charged with building the Palace. Barry supervised the main design, layout, and management; Augustus Pugin assisted with the detailed design and elaborate decoration.

Barry decided that a clock tower with a *King of Clocks* would be an important part of the design. He commissioned the clockmaker to the Queen, Benjamin Lewis Vulliamy, in 1844 to produce a design for the clock. However, the Commissioners for Works decided that the clock was too important for Barry to manage himself, so they wanted the clock's manufacture to be put out to tender. George Biddell Airy, the Astronomer Royal, was engaged as referee: he produced a specification for the clock, including the requirement that the first blow on the hour bell should be accurate to within a second of time. Clockmakers of the day dismissed his requirement as impossible to attain. Edmund Beckett Denison, a young lawyer and amateur horologist, who had written a book on turret clocks, was also engaged as a referee.

Denison designed a new clock, different from Vulliamy's design, and its construction was put out to tender. In 1852, the company of Edward John Dent won the contract; Dent was a maker of watches and chronometers and had only just started to make turret clocks. Established makers of turret clocks were dismayed that Dent had been chosen.

By 1858, the clock tower was well enough advanced that the Commissioners decided to

FIG 1.2 Edmund Beckett Denison, QC, designer of the clock and bells.

order bells for the tower; there were to be four quarter bells and a large hour bell of about 14 tons. Denison was the main referee and he produced a design and specification for the bells' sizes and shapes, and the composition of the bell metal. Warners cast the hour bell at Norton on Tees, it was shipped to London, and the great hour bell that weighed over 16 tons was put on display and sounded every day with a large hammer. It acquired the name *Big Ben*, after Sir Benjamin Hall, who was the Chief Commissioner for Works of the day. One day, the bell cracked disastrously and the bell had to be broken up and recast, this time by the Whitechapel Bell Foundry.

In 1858, the bells were hauled up the almost-finished tower and installed in the belfry. The clock was installed and set going the following year. After a few months' operation, it was discovered that the Great Bell was again cracked, so it was taken out of service and the hour struck on the largest of the quarter bells. Big Ben was not used for several years. Airy then proposed that the bell be turned so that the crack was at a point of minimum vibration. This was done and the bell has been striking successfully ever since, despite two surface cracks.

In the years that followed the Great Clock performed well, keeping time to better than a second a day for about 80% of the year. Its excellent performance was mainly achieved by the escapement; the device that linked the pendulum to the clock train. Denison understood that the effect of wind and weather on dials was transferred to the clock mechanism and on to the pendulum, causing variations in timekeeping. He invented the double three-legged escapement, which isolated the pendulum from the effect of

FIG 1.3 The great bell 'Big Ben'. This figure is reproduced in colour in the colour plate section.

FIG 1.4 The Great Clock movement.

5 *LONDON S. W. — The Houses of Parliament. — LL.*

FIG 1.5 The New Westminster Palace about 1905. The clear view unmasked by today's trees shows the general
layout and how the Hall was integrated into the New Palace.

the weather on the dials. This escapement was never patented and was used on all high-quality tower clocks from the Westminster Clock onwards. A compensation pendulum was also used to eliminate the effect of varying temperature on the timekeeping. Denison also popularized the flatbed construction of turret clocks, which allowed for easy assembly and servicing. Many church and tower clocks made after 1860 followed this pattern and also incorporated the Westminster Chimes, as they became known.

There were various stoppages of the clock, for example, due to freezing snow, flocks of birds sitting on the hands, and a ladder being left against the mechanism. From time to time, the clock has been serviced and out of action for a short time. Servicing was carried out by the makers, Dent, until 1974 when Thwaites and Reed took on the task. In 2004, maintenance of the clock was taken in-house.

In 1976, there was a major disaster: the whole of the quarter-striking mechanism was almost totally destroyed. A shaft had broken, owing to fatigue and the driving weight rapidly descended the height of the tower, causing the damage. Thwaites and Reed masterminded the rebuilding, which was completed in time for the visit of the Queen to the Houses of Parliament.

The clock underwent a major overhaul in 2007 to correct some severe wear on the striking train and going trains.

The 22½-ft diameter dials are around 200 feet above ground level. Big Ben weighs 13½ tons and the clock mechanism is 14 feet long. Resident clockmakers climb the 290 stairs to the Clock Room three times a week to wind the clock.

Big Ben remains a reliable mechanism, telling time with an unparalleled accuracy for a mechanical timekeeper. Long may Big Ben continue to serve and represent the British nation!

CHAPTER TWO
A VISIT TO THE CLOCK TOWER

THE PALACE CLOCK TEAM

We are privileged to have a special tour by kind permission of The Keeper of the Great Clock. As well as looking after the Great Clock, he is responsible for every aspect of maintenance around the Palace and has a diverse team of craftsmen including glaziers, locksmiths, carpenters, masons, and, of course, some clockmakers. The Palace has over 2,000 clocks in its many rooms; these include mechanical dial clocks, long-case, and battery-driven clocks, as well as two systems of master clocks. Needless to say,

there are some very fine clocks at Westminster. A team of three clockmakers look after all the clocks in the Palace; their day starts very early with a winding round that has to be completed before Palace staff move into their offices. If, when it is being wound, a clock is found to be needing attention it is taken back to the workshop for repair and testing before being reinstated.

The Great Clock has to be wound three times a week, Monday, Wednesday, and Friday. Twenty-four-hour cover is provided by the team of clockmakers to cope with the unlikely event of a problem; one clockmaker has to be within a two-hour journey to the Palace at all times.

FIG 2.1 The Houses of Parliament with the Clock Tower in the foreground. Portcullis House is the modern building to the right.

OUR PRIVATE TOUR

Since not everyone can easily travel to the Clock Tower, here is an illustrated tour that anyone can join, and all without climbing the 334 steps leading to the belfry some 220 feet above street level. Since our tour is a special one it will take us into areas where it is not possible for the public to visit for reasons of time, security, and safety.

Entry

Our journey starts at the bottom of the Clock Tower in the south-west corner through a door marked, not surprisingly, *Clock Tower*. This takes us into a small rectangular lobby with a spiral staircase protected by an iron handrail. Progress is afforded with some relief of the monotony; there is a small landing after every dozen steps or so.

For the first 70 feet, there are no windows in the stairwell, since on the south side the tower is attached to the House of Commons building and on the west side the windows must have been omitted to make the union of the tower with the building stronger. However, we see some window-like openings on the wall of the air shaft. Initially these were windows and would have let a little light into the stairwell but they have now been bricked up. Soon after starting building the Clock Tower, the planned use of the air shaft changed from taking air in at the top to using it as a vent. A fire was to be used to make the air rise, so the windows had to be bricked up. After a while, windows appear in two sides of the staircase. These give views over New Palace Yard towards Parliament Square and over the roof of the House of Commons chamber.

FIG 2.2 The Clock Tower from New Palace Yard. The entrance to the tower is through a passage between the tower and the main building.

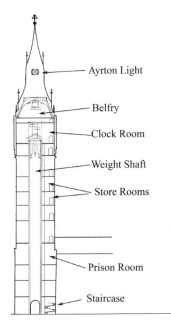

Ayrton Light

Belfry

Clock Room

Weight Shaft

Store Rooms

Prison Room

Staircase

FIG 2.3 A sectional view of the Clock Tower, looking east.

Facing north with windows on our left and behind us, the wall in front of us is just like any other wall. However, behind the brickwork is an air shaft that runs from near the bottom of the tower all the way up to the belfry. It was an early attempt at air conditioning, but it was a failure.

The First Visitor Room

After climbing a great many steps and passing one door we arrive at Room 1. In the past this has been referred to as the Prison Room. However, the real Prison Room, though it is in the Clock Tower, is only accessible from the House of Commons, not from the Clock Tower staircase. The Prison Room was designated as a room where errant MPs could be locked up if necessary. The last time it was used was in 1880, when the MP Charles Bradlaugh was imprisoned here for refusing to swear the oath of allegiance to Queen Victoria. No doubt this room was one of the nicest cells a prisoner could occupy, since it is oak panelled and had a bedroom and sitting room. An old engraving depicts the Prison Room in all its luxury and its accuracy is confirmed by original architect's drawings.

On entering we find that the room is an odd shape with what seems like a large chimney breast that dominates the room. This is formed by the walls of the weight shaft, down which the clock weights descend; it runs all the way from the clock room to the ground floor. An alcove faces the chimney breast on the right. Behind the weight shaft and the wall of this alcove runs the air shaft mentioned earlier.

Benches provide a welcome resting place for visitors. Today, various items are displayed. A cabinet holds a selection of parts: a spare suspension spring for the pendulum, a massive two-handed spanner for removing the octagonal nuts that retain the minute hands, some oilcans, and

FIG 2.4 The entrance to the Clock Tower.

FIG 2.5 Looking up the stairwell. There are 334 steps to the belfry.

FIG **2.6** Engraving of the Prison Room, from *The Illustrated London News*, July 1880.

FIG **2.7** Looking out of Room 1 to the stairwell, showing the display case.

FIG **2.8** An oilcan, minute nut spanner, suspension spring, and timing record inside the display case.

other items. The windows are tall and narrow: on the windowsills are two artefacts connected with the bell, Big Ben. One is the massive 6½ cwt hammer head that originally tolled the great bell. When the bell cracked, the hammer head was made lighter. The original was put on display on the River Terrace but has since been moved to the first visitor room. Another is the ball from the clapper that was installed inside

Big Ben. No provision was ever made to swing the Great Bell or to pull the clapper, so the clapper was removed in 1956 when the bell hangings were refurbished.

Room 4

Continuing our ascent, we reach Room 4, which is now open as another visitor room. Again there is time to rest and we can see computer

FIG 2.9 Clock hammer head and bell clapper ball on display.

FIG 2.10 Old patterns for Great Clock wheels.

animations of the Great Clock mechanism. The escapement, chiming, and hour-striking can all be seen. A model of the Great Clock's escapement can also be inspected.

On display here are some of the original wooden patterns used to make cast-iron parts for the Great Clock. A pattern is a wooden model used to make the appropriate part in iron. A full-size pattern was placed in a special two-part iron box and filled with sand that was rammed down hard. Then the box was split into two and the pattern carefully removed, leaving a cavity the same shape as the required part. The cavity was then filled with molten cast iron and, when cool, the iron part was removed and, if necessary, machined. It is quite amazing that these patterns have survived so long after the clock was completed; no doubt the maker was required to give up all the patterns made.

In 1976, some major work had to be done on the clock; a wooden pattern and a steel wheel made from it at that time hangs on the wall. This particular wheel was found to have a flaw in the casting, so was rejected.

The Dials

From the visitor room we move upwards, passing numbered doors that give entry to the various store rooms. Looking down, the spiral staircase now vanishes into the gloom beneath us.

With some trepidation, we hear that the clock room is number 9: a long way to go yet. We continue to pass doors, all in the same position. Then, for those with a good sense of direction, door number 8 appears almost opposite to where it is expected; its position is such that it

FIG 2.11 Looking down the stairwell. The rope is used for hauling equipment up the tower.

looks like it would take us outside the tower! This is the access to the space inside the dials. Passing through, we are relieved to find that we are still inside the tower and behind one of the gigantic 22½ ft diameter dials. Since the dials project out from the clock tower we had to pass through the structural wall of the Clock Tower to get to the space behind the dials; hence the access door apparently leading to the outside of the tower.

A passageway passes along behind the dial, through the corner pillar and on to the next dial. It is possible to walk round behind all four dials and end up back at the entrance door. Stopping to look at one dial, we see that the frame is cast iron that has been made in sections and bolted together. Opal glass has been used for the glazing: this is a glass that is translucent and milky white in colour, giving a good appearance from the outside, allowing light through during the day and an even illumination at night. Each dial has a total of 312 separate pieces of glass. The sun throws a shadow of the minute hand onto the glass. Since a minute mark is a foot long we can clearly see the hand advance every two seconds: the phrase *Time Flies* now takes on a real meaning.

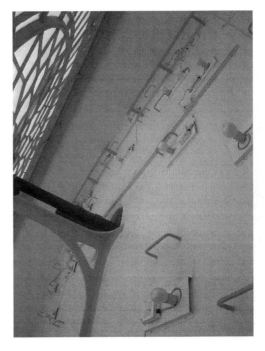

FIG **2.13** Dial illumination.

FIG **2.12** Behind a dial.

FIG **2.14** Access to the next dial.

FIG **2.15** The hand's shadow shows 23½ minutes past the hour.

FIG **2.16** Dial with inscription, probably devised by Pugin. This figure is reproduced in colour in the colour plate section.

dials we would see a Latin inscription that runs below each of the clock dials *DOMINE SALVAM FAC REGINAM NOSTRAM VICTORIAM PRIMAM*, which means *O Lord, keep safe our Queen Victoria the First.*

Room 9: The Clock Room

Arrangement

Looking upwards, we see a large bracket that supports the drive to the hands.

Turning round with our backs to the dial we see the main tower wall, since the dials stand proud of the tower. A couple of lancet windows are above; these provide some light into the clock room. Iron rungs are let into the wall, enabling a person to climb anywhere; originally these gave access to the multitude of gas jets that were used to illuminate the dial. Today, light is provided on each dial by 28 high-efficiency 85 watt bulbs. Poking out of the floor is the blocked off-end of a pipe, the remains of the old gas supply.

We are now about 200 feet above where we started, if we were able to look outside at the

Leaving the dials, we climb a few steps more and arrive at last at Room 9, the clock room. We find the clock movement immediately on our left. Moving round to the front of the clock, we can inspect the general layout. Above the clock there are two massive two-bladed fans called flies, one for the hour-striking train, and one for the quarter-striking train. These flies regulate the speed of the chiming and striking trains. Two large iron girders run east–west across the room. These support the drives to the dials and a lifting winch that was used to install the clock in 1859.

Looking now at the clock in detail, we see that the frame is 14 feet long and the wheel work mounted on this is in three sections. Each end of the frame rests on a pillar; these are of brick

FIG 2.17 The Great Clock from the front.

and, in fact, are the very tops of the weight shaft walls that go all the way down to the tower foundations.

Along the bottom of the frame of the clock runs the inscription, '*This clock was made in the year of our Lord 1854 by Frederick Dent, of the Strand and the Royal Exchange, Clockmaker to the Queen,*

FIG 2.18 The Great Clock, showing the flies above and the connections to the dials. This figure is reproduced in colour in the colour plate section.

from the designs of Edmund Beckett Denison QC'. In the centre is an oval plaque that records, '*Fixed Here 1859*'.

In the middle of the clock movement is the going train, which tells the time and drives the dials. A quarter-striking or chiming train on the right sounds the quarter chimes every 15 minutes, while the hour-striking train on the left sounds the hours.

Going Train

We start our tour of the clock with the going train in the centre. Motive power comes from a weight of about 2½ cwt that is suspended on a steel wire line. The line passes round a barrel and through a train of gears; the descending weight turns the hands of the dials. The pendulum is connected to the clock by an escapement that allows one tooth of the escape wheel to advance for every beat of the pendulum. In return, the clock gives the pendulum a small impulse to keep it swinging. For the Westminster clock,

Edmund Beckett Denison employed his double three-legged gravity escapement. This is a unique design, and it isolates the pendulum from variations in power caused by different weather conditions. The end result is the astonishingly good performance of the clock. The pendulum hangs from a large bracket that is fixed to the wall. Only the top of the pendulum is in the clock room; the main part of the pendulum is in a compartment below the clock.

On the front of the clock are two dials. A white-painted one tells the minutes and a brass one tells the hours from 1 to 24. Today this dial serves no purpose and is a legacy from the past, when it was planned to use the clock to turn the dial gas illumination on and off. A third dial, also of brass, is mounted inside the clock; this gives the time exact to the second. It has a single hand that turns once in two minutes, so the dial is engraved from 5 to 60 twice. Additionally, the clock has two centre arbors that turn once each hour. On one is mounted a four-lobed cam that releases the quarter chimes every 15 minutes; the other has a single cam that is part of the hour-striking release system.

We can now see how the hands on each dial are driven. A large bevel gear on the centre arbor drives a similar bevel gear on a shaft that ascends at an angle. Above the clock, this shaft engages through more bevel gears with the drive shafts that go off to each of the dials. These are known as leading-off rods and turn once an hour to drive the minute hands. Behind each dial and inside the clock room are sets of motion works; these form the gearings that provide a 12 to 1 reduction to drive the hour hands. A Y-shaped frame with two weights can be seen at the front of each set of motionworks; this is a counterbalance for the minute hand and sets it in perfect balance, so the clock needs minimal force to drive the hands. If there was no counterbalance, the clock would stop at quarter to the hour not having enough power to raise the hand. The hour hands too have counterbalances, but these are rather hidden from view behind the motion-work gearing.

Also on the beams above the clock and the bevel gears is a double-barrelled crab; a lifting winch that was used to raise all the clock parts up into the clock room.

FIG 2.19 The going train.

FIG 2.20 Leading-off work; the connection to the hands.

FIG 2.21 Motionwork showing counterbalance
for minute hand.

FIG 2.22 The quarter-striking train.

Quarter-striking

To the right of the going train is the quarter-striking train. On this barrel, the driving weight is about 1 ton. A fly is driven through a train of gears. This operates as an air brake and, as the vanes beat the air, the train is made to run at a steady speed. On the great wheel there are five sets of cams; these raise the bell hammers at the appropriate time and then release them, thus sounding the bell. There are four bells, but an additional set of cams work a second hammer on the fourth bell since a note is struck twice in succession and the clock would not be able to operate a single bell hammer twice with sufficient speed.

When the clock was being built, it was realised that there was not enough space in the clock room to house the flies at the back of the clock. Normally, on turret clocks the flies are mounted behind the clock movement. At Westminster, it was found necessary to build the clock with the flies mounted vertically above the clock, driving them by bevel gears.

On the front of the quarter train is a wheel with slots, known as a count wheel. The slots are at different positions and allow the train to run a sufficient length for one, two, three, or four quarters. In fact, the sequence is repeated three times, so the count wheel only rotates once in three hours.

Since the driving weight is very heavy, the clock was originally designed with reduction gears so that the clock could be wound by hand. At the extreme right of the base frame, there is a winding handle; through two winding wheels, the effort is reduced making it possible to wind such a large weight. When it was wound manually, the winding wheel had to be engaged. Then just over 128 turns on the winding handle produced one turn on the barrel. The same system is used on the hour-striking train.

Hour-striking

The hour-striking train is very similar in construction to the quarter-striking; it is placed to the left of the going train. Instead of rings of cams, there is only one set placed on the back of the great wheel. The arrangement for striking the hours is similar to the quarters. On the great wheel are ten large cams that lift and drop the hammer for Big Ben. On the front of the clock

FIG 2.23 The hour-striking train.

FIG 2.24 The cams that release the quarter-striking.

is another count wheel; this time the notches are spaced at increasing intervals and these determine the number of blows to be struck each hour. The count wheel turns once in twelve hours. Above the hour-striking train is another vertical fly that regulates the speed of striking.

The Clock Strikes

As the hour approaches, we can see the whole sequence of the clock striking. The first stage of the release of both the quarters and the hour is known as the warning. The cam with four lobes on the centre arbor lifts the locking lever on the quarter train until the train is released and warns, advancing to the second position of locking. When the lever drops off the cam, the train is then free to run and the quarters are sounded. The train runs, turning the count wheel until the locking lever drops back into a slot on the count wheel and the train is locked.

First, the quarter train warns with a smart click just before the hour followed by the hour train. The great fly above turns slightly in preparation. At about 20 seconds to the hour, the quarter train is released with a clunk, the release lever drops, the train starts running and the fly

above churns the air. In the belfry above, we hear the bells striking, though somewhat muffled. There is a lot of rattling from the linkages that connect with the bells and, as the last note sounds, the train locks, and the fly carries on turning and freewheels, the clicks work on the ratchet, making a loud noise like a football rattle until the fly ceases to turn. All is quiet, except the regular beating of the clock escapement. After about 5 seconds, at two seconds to the hour, the striking train is released with a bang, which then starts to run, lifting the bell hammer the last necessary fraction and letting it drop with a resounding note from Big Ben aloft. Blow after blow sounds and then, when the hour is complete, we are treated to another ear-shattering clatter as the striking fly spins to a halt.

1976 Disaster

In 1976, the fly arbor on the quarter-striking train failed, owing to a fatigue fracture. With no speed regulation, the 1 ton weight descended rapidly until it hit the bottom of the weight shaft. The consequence of the accident was that all of the quarter-striking train was destroyed, with almost everything broken into small parts. Fortunately, little damage was sustained to the going train and hour-striking. The whole quarter-striking train was rebuilt and the clock was set going again in 1977, in time for a visit to Parliament by Her Majesty the Queen.

Looking at the clock today, there is little evidence of the disaster, but close inspection shows that on the quarter gears the tooth form is now of a modern design. Just below the words 'Clockmaker to the Queen' in the inscription that runs along the bottom of the clock there is a post on the floor supporting the frame. This has been positioned where the frame broke and provides extra support to the repaired structure.

Behind the Clock

As part of this special tour, we are allowed to see behind the clock. Climbing up a small ladder to the right of the quarter train, we go over the top wall of the weight shaft and step down behind the clock. Here another setting dial can be seen; this is used for setting the hands to time in the event of the clock having to be stopped for maintenance.

Turning away from the setting dial, we open a trap door in the steel floor and, after turning

on an inspection lamp, descend an iron ladder to inspect the pendulum. We find ourselves in a small box-like compartment: calling it a room would be an exaggeration. It is made of an iron sheet on one side and has the brick walls of the weight shaft on the others. This compartment provides a draught-proof compartment for the huge pendulum that swings there. The pendulum bob weighs 700 lb; this is equivalent to three good-sized men. Underneath the bob is a big nut. This is used to adjust the pendulum; raising the bob makes the clock gain, lowering it make the clock lose. There is also a plate with graduations to measure the arc of the pendulum; we see the bob swings out to 3° on each side of the zero position. An iron tube supports the bob. In fact, although we cannot see them, there are

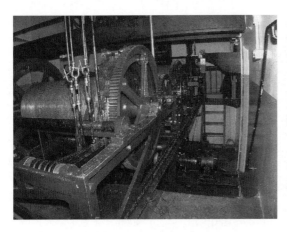

FIG 2.25 Behind the clock.

FIG 2.26 The bob of the pendulum.

concentric tubes of iron and zinc, which together compensate for changes in temperature so effectively that the pendulum stays the same length. Holes in the tube allow air in and this means that if the temperature changes suddenly then the compensation system can respond faster. Having carefully avoided touching the pendulum, we leave the pendulum pit and close the trap behind us.

Turning our attention now to the top of the pendulum we see that is suspended by a thin spring that hangs from a massive cast-iron bracket let into the wall of the air shaft. The pendulum rod begins as a rod and then soon there is a collar where the assembly of compensation tubes starts. This collar carries a multitude of pieces of metal, including some old penny coins. Fine regulation of the clock is still done with old, pre-decimal, pennies. Adding an old penny causes the clock to gain 2/5 of a second a day. Removing a penny causes the clock to lose.

There is a good view of Denison's double three-legged gravity escapement. A pivoted arm is positioned on each side of the pendulum.

FIG 2.28 Denison's double three-legged gravity escapement. This figure is reproduced in colour in the colour plate section.

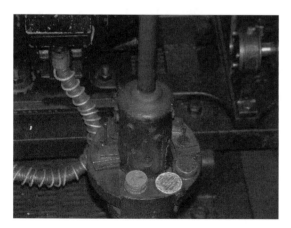

FIG 2.27 Add one penny and the clock will gain 2/5 of a second per day; taking a coin off makes the clock lose. This figure is reproduced in colour in the colour plate section.

As the pendulum swings away from us the far arm is lifted. This releases the train and we can see that the near arm is reset or raised a small amount. As the pendulum swings back towards us, the far arm descends, giving an impulse to the pendulum. The near arm is now lifted and we see the far arm being reset.

Returning to the quarter train, we get a good view of the hammer-lifting cams; the size of the machinery is truly that of a steam engine. Looking down into the frame, a modern-looking piece of equipment can be seen spanning the lower part of the winding wheel. This is a safety brake that was designed by the National Physical Laboratory. In the event of the quarter barrel over-speeding, the brake engages and stops the wheel from turning. An identical brake is installed on the hour-striking train and with these brakes installed, there would be no damage caused to the clock if any part of the striking trains failed.

Passing through what appears to be a narrow window opening in the wall behind the clock, we come into a tiny compartment. In here is the motionwork for the west dial. This unusual arrangement is made necessary by the air shaft that passes up through the tower. Between our feet and the ground floor 200 feet below there is just a sheet of iron.

Coming out of the compartment, climbing over the weight shaft wall, and passing round the front of the clock, we can then climb down behind the striking train. Here we see the rear of the striking great wheel. Large cams cast onto the back of the great wheel lift the hammer for Big Ben. We can also see the rear of the seconds dial, which has been made with two sides so that it can also be read from the rear of the clock.

Winding the Clock

With the hour just struck it's time to wind the clock. An electric motor installed below the movement connects with the quarter and hour-striking barrels and this relieves the team of some long and hard work, but the going train is still wound manually. The team of clockmakers demonstrate how this is done but first we are handed ear defenders. The motor is started, which gives a deep vibration to the floor and, with the clutches thrown in, winding commences. Each of the huge striking barrels has six clicks: these fall with a resounding clack, clack, clack. We are indeed grateful for the ear defenders. It takes about 40 minutes to raise the two weights; there is 1 ton on the quarter train and 1 ton on the striking train. The going train, by comparison,

FIG 2.29 Winding the going train by hand. This figure is reproduced in colour in the colour plate section.

has a mere 2½ cwt, and this is wound manually. With all the weights fully wound, the motor is shut off and we can remove the defenders and return to normal conversation.

There are two events in the year when the Great Clock has to be absolutely spot-on time: Remembrance Sunday and New Year. On Remembrance Day, the striking of Big Ben has to coincide exactly with the firing of guns to signify the commencement of the two minutes silence. The New Year is heralded by the chimes, which are broadcast across the world.

Plaques

On the wall is a painted plaque that records in gothic script:

WORDS AND MUSIC OF THE CHIMES:
ALL THROUGH THIS HOUR
LORD BE MY GUIDE
THAT BY THY POWER
NO FOOT SHALL SLIDE
Music from Handel's Messiah

On a carved wooden plaque below this runs the legend:

1859–1959
This stone commemorates the Centenary of the Bell and
Great Clock of Westminster

A small notice explains that the wooden plaque is a copy of a carved stone that was placed at the base of the tower. The original stone can be seen at the base of the tower from Bridge Street.

The Belfry

Taking the stairs upwards from the clock room we finally come to the top and see door number 10, which leads out to the belfry. In the last turn we have to duck under a stout tie rod that crosses our path just a little under head height. Upon entering the belfry, we stop for a few minutes to take our breath and to admire the view through the columns that support the lantern roof above us and allow the sound of the chimes to escape. Facing north, we look down the river over White-hall; to the east we see London south of the Thames. To the south the whole expanse of the roofs of the Houses of Parliament lie below us and, despite our elevation, the Victoria Tower at the far end of the Palace towers above us. Looking west affords a good view of Parliament Square, St Margaret's

FIG 2.30 Looking east over the river with the London Eye just in view.

FIG 2.31 The great bell: Big Ben.

church, Westminster Abbey with St. James Park, and Buckingham Palace in the distance.

The great bell, Big Ben, takes up the centre of the tower; it certainly is a monster. A rail separates us from the bell; no one would want to be hit by the 2½ cwt hammer head when it falls to sound the hour. At just over 13½ tons the bell is 7 feet 6 inches high and 9 feet in diameter; it is an agreeable grey-green colour, the result of 150 years of weathering. (The bell was blacked in 2009 as part of the clock's 150th anniversary celebrations.) A large wooden platform under Big Ben is only a foot beneath it. This was installed to catch the bell, should its cracks ever develop suddenly so that it falls apart.

The construction of the bell frame is interesting. Sixteen cast-iron standards rise from the floor to support the bell frame, four on each side. In construction, the frame is a square of wrought-iron beams with two more beams crossing in the middle. The hour bell, Big Ben, hangs from the centre of the cross and the four quarter bells are hung on short diagonal beams, one at each corner of the frame. Massive corner braces stiffen the frame, which was found to be far too weak when it was first used. A mass of rivets hold the beams together making it look more like an ancient ocean liner than a bell frame. Each bell has a hammer that is pivoted at the top of the bell, there is one arm that sweeps down and terminates in a hammer head; the other extends horizontally and is pulled down by a steel cable that vanishes into the floor below. To prevent the hammer head chattering on the bell, a fixed rubber pad allows the hammer to strike the bell and then to be kept just clear of the bell so that it can sound properly. Each bell has a top shaped like a button or mushroom, which is gripped by an iron collar. Although quite unconventional for bells, this method of hanging has been very useful at Westminster, since it has enabled the bells to be turned when their surface has been worn by the repeated blows of the hammer.

Passing through the gate, we can inspect the bell close up and see the moulding wires, tracery decoration, and the inscriptions on Big Ben. Looking at where the weight of the bell appears, we see that this part of the inscription was cast as a block and the figures chiselled out after the bell had been weighed; the match of characters is very good. There are two coats of arms cast on opposite sides of the waist of the bell, the Royal Arms and the Arms of the Palace of Westminster, the portcullis that Barry the architect had adopted as his insignia.

Within a few weeks of the clock and all the bells being set going, it was found that Big Ben was cracked. Moving round to where the cracks are we see a large square-section cavity about six inches square and four deep. A crack runs upwards from the excavation and another down to the lip of the bell. After the crack was discovered in 1859, metal was excavated to discover just how deep the crack went. Two feet further on round the bell, another shorter crack can be seen. The cracks have not moved in living memory but they are monitored regularly as a safety measure. Big Ben was turned after it cracked to bring the crack to an area of minimum vibration and a lighter hammer head was installed.

Returning to Big Ben's hammer, we see that its head is quite a way from the bell and is not resting on its rubber pad. The hammer has to be 'left on the lift' in order to get the first blow of the hour accurate to within a second of time. The striking train is released a few seconds before the hour, then the hammer is raised a fraction more before falling to sound the bell. The hammer is again raised by the clock and falls six seconds later, and so on until the full hour has been struck. After the last blow, the hammer is raised almost to the point of dropping to be ready for the next hour.

On coming out of the bell enclosure, we look towards New Palace Yard; our progress is barred by a railing. Looking downwards, we see that the flooring is a cast-iron grating; once we focus our eyes through the grating we see that we are looking down the inside of the great air shaft, which descends out of sight. Sunlight filters in

FIG 2.32 The hammer head is seen well away from the bell.

FIG **2.33** The top of the air shaft.

through the narrow slit windows to give a great sense of the size of the shaft and the tower.

Getting on our hands and knees we look under the wooden platform below Big Ben. A rusty iron hammer meets our eyes. This is the hammer that was fitted to the fourth quarter bell after Big Ben had cracked and before it was quarter turned. This meant that the fourth bell once had three hammers as, for several years, it served as the hour bell as well.

From the belfry, our extended tour takes us up to a walkway that goes round the inside of the tower above the level of the bells. Looking down on the top we see how each bell is supported from an iron plate. We also see clearly that the fourth bell has two hammers opposite each other; this is so that two notes can be sounded in quick succession.

A collection of microphones in a corner are secured to a support girder. These are used by the BBC to broadcast the Big Ben chimes live.

Another iron staircase spirals upwards. In the next colonnade up is the Ayrton light, which is lit when either House sits at night. This was introduced in 1873 by First Commissioner Ayrton and originally only faced towards Buckingham

FIG **2.34** The fourth quarter bell.

Palace, so that Queen Victoria could know when her Parliament was at work; today its light is omnidirectional, so that all can see it.

On returning to the belfry, those with sensitive hearing insert some foam ear plugs. It is a minute to the hour. At about 40 seconds past the 59th minute, the quarter chimes start; the sound of the four phrases is majestic and somewhat loud, to say the least. A period of silence follows, emphasized by the dying hum of the last bell. We then see Big Ben's hammer raise a fraction, then fall to give a powerful note just on the hour. Again, the hammer rises like an automaton, and again strikes the bell. The hour is measured out and we not only hear the bell but feel it as well; the note is deep and wild and is accompanied by a pronounced beat. Today a bell founder would

FIG 2.35 Looking up inside the spire.

The Link Room

Coming down from the belfry, we espy a door that is in a somewhat awkward position; this was once called the Lever Room but is now known as the Link Room. Unlike other doors it is not numbered. Once the door is unlocked, we enter by hopping across the stairwell. *Mind your heads* is a timely warning, since the room is only about 5 feet 6 inches high. Walking around with bowed shoulders, we see how wires come up from the clock in the room below and their action is then transferred across the room by bell cranks, which are levers that turn the action through 90°. There is a set of cranks above the quarter train and links at other positions below the appropriate bells. We see the top bearings of the flies in the floor. There is no bell crank for Big Ben; the wire cable goes straight from the clock below to the bell above.

Looking upwards at the roof, we see that it is vaulted and made of brick. The whole tower is fireproof; the only combustible material used is the wood for the doors to the various rooms. Looking around, we see tie bars that run all the

be distressed if a newly cast bell came out with such a strident beat note, but there it is and it is part of the sound of the National Icon. With the last note sounded, the hammer lifts as if to strike another note, but it stops when fully raised; this is only the hammer preparing to strike at the next hour. Ben's note carries on humming; we can still detect him speaking a good minute after the last blow.

The Westminster quarters were taken from the chimes of St. Mary's church in Cambridge. A Dr. William Crotch (1775–1847) is attributed with the tune and, as the board in the clock room records, it was based on a phrase in Handel's *Messiah*.

The visit to the belfry is now complete and we head downwards to see all the other rooms.

FIG 2.36 Inside the link room. The cluster of cranks is directly above the quarter train and the bar on the floor to the left holds the fly arbor bearings.

way round, and across the room. The iron standards that support the bell frame rest on the tower walls; the tie rods stop the tower bulging out as a result of the weight of the frame with its bells, something approaching 40 tons.

Negotiating a difficult step out of the link room, we now head downwards, passing the door to the clock room and the entrance to the dials.

Weight Shaft

We reach another door, number 7. Upon entering, we see that we are inside the room directly below the dials. Unlike the other rooms, this room has very short windows and a lower ceiling. Peering out of a window, we see that the tower walls appear to be very thick here. In fact, the tower here starts to swell out just below this level, so we can see the overhanging projection of the dials above. The room follows the layout of the others, with the exception that in the middle of the weight shaft is a door bearing a warning notice:

DANGER
VOID BEHIND THIS DOOR

FIG 2.37 Weights can just be made out in the weight shaft below.

On being unlocked, the door reveals a dark hole: a safety bar across the entrance prevents us from inadvertently stepping down the weight shaft. Since they have just been wound, the three clock weights are opposite us. Each is hung on a massive pulley about two feet in diameter; their steel wire lines are made off under the clock. From its tie off point, each line then passes down, round its pulley, and back up, where it is wound onto a barrel on the clock.

With the aid of a powerful torch, the shaft is made out to be constructed of brick, vanishing into nothing in the gloom far below. At the top of the weight shaft is a ceiling formed by the iron grating that is the floor immediately underneath the clock; here the standing ends of the weight lines are made off. Above, on the opposite side of the shaft is the iron box-like compartment that contains the pendulum.

It was through this weight shaft that Big Ben was raised up to the belfry. Having seen the giant bell, somehow the shaft just does not seem big enough to have been able take the monster up to his final home above London. Once Big Ben was raised, the other bells followed and, finally, the clock, in pieces, was hauled up the shaft.

With the weight shaft access door safely locked, we exit Room 7 and head downwards. It is nice to be going down, much easier work than coming up.

All the Other Rooms

Visits to the other rooms reveal a collection of miscellaneous items related to the clock. Rather than a blow-by-blow account of what is in each room, let us look at what is being stored in general.

Three types of item are preserved in the store rooms: old original patterns used in the making of the clock, broken parts from the clock following the 1976 disaster, and new patterns used for the repairs in 1976. Some large specially made wooden boxes of unusual proportions contain patterns and the larger broken parts; steel boxes of various sizes hold the smaller broken parts.

Some patterns are easily identified. One is for the pendulum bob, complete with its core prints. To produce a round hole through the bob, a cylinder of sand, called a core, was inserted in the mould; this saved a lot of time and effort in producing a hole rather than having to use a drill after the bob was cast. To hold the core in place, the pattern has round extensions on each end of the wooden bob. When the mould for the bob was made, the tube of sand was inserted into the core prints. Another pattern is for the great wheel for the striking train; its ten cams for lifting the striking hammer clearly identify what the pattern is. Teeth on the wheel have been made of wood, carefully profiled, and attached to the periphery of the wheel in pairs. Modern patterns are painted in bright colours and include a large section of the base frame, a great wheel, and many other smaller items needed for the restoration.

Broken parts are stored in abundance; they range from a very heavy section of the main frame through broken wheels to the odd teeth or cams that were shattered when the clock failed. One item is the fly arbor that failed. This is a wrought-iron gas pipe about 1¼ inches in diameter and was made by folding a strip of iron round a mandrel and heating the joint before hammering it to produce a weld. A mechanical hammer must have been used, since the joint shows up as a neat depression all the way along the length of the arbor.

It is indeed a credit to British engineering that the clock was repaired quickly using traditional and appropriate techniques.

FIG 2.38 Numerous parts broken in 1976 are stored and labelled.

The Real Room 1

Passing by Room 1, we continue downwards until we find an unmarked door. Behind this, we can see not a room, but a space more like a storage cupboard. Fixed to the wall is a small rack of electronics units with glowing lights. This is the interface between the microphones in the belfry and the telephone landline that transmits the chimes of Big Ben to the BBC. On the wall is an ancient gas light, the burner with its mantle long since missing. Somewhat artistically, an old set of crystal-set-type headphones hang on the gas light, a sort of tribute to an earlier installation. Could they perhaps date back to 31 December 1923, when the midnight chimes of Big Ben were first broadcast by the BBC to welcome the New Year?

A square wooden access hatch in the wall reveals an entrance to the bottom of the air shaft. The whole atmosphere is somewhat eerie and deserted, the shaft soars upwards as far as we can see, lit by the tall lancet windows. Access to the weight shaft is also possible from this point.

On leaving the electronics room and returning to ground level, we see two marks on the floor of the stairwell. A brass button in the centre represents where the builder's plumb bob had to fall when the tower was being built. Some 8 inches away is the present datum, indicating that the tower has a slight lean. When the Westminster section of the underground Jubilee line was constructed between 1995 and 1997, thousands of tons of concrete were pumped into the ground to ensure that the tower did not lean any further under the influence of tunnelling work.

The top of a large stopcock partially protruding from the wall is noticeable. It is believed that this was the main gas supply to the tower; it would have controlled the illumination to the dials and the Ayrton light when they were lit by gas. The supply might have been turned off during the day and then turned on when night fell. Gas installations like this often employed a bypass that allowed

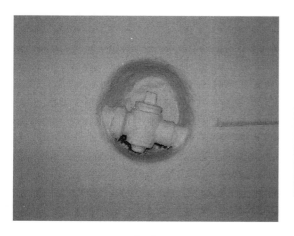

FIG **2.39** Main gas tap.

FIG **2.41** New Palace Yard. The bell Big Ben was taken through the large doors on the left to raise it up the Clock Tower.

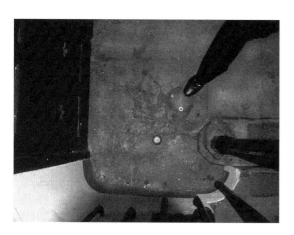

FIG **2.40** Datum marks for the builder's and today's plumb bobs.

FIG **2.42** Normally the weights are out of view, but here they can be seen resting on the pile of sandbags during the 2007 restoration.

FIG **2.43** A final look.

THE GUARDIAN OF THE CLOCK TOWER

FIG **2.44** The Guardian of the Clock Tower, from *The Graphic* of October 1887.

a small amount of gas to flow keeping the burners alight at a very low level throughout the day.

Immediately outside the entrance to the Clock Tower stairs is a door that gives access to the bottom of the weight shaft. This is no longer in use, since the base of the shaft has been filled up with sandbags, a precaution to soften the fall of a weight in the unlikely event of one of the lines breaking.

Our final port of call is the bottom of the weight shaft. We return to the quadrangle of New Palace Yard; a spectacularly long key unlocks the massive double doors at the bottom of the tower that face onto the courtyard. On entering, we find ourselves in a plant room full of long cylindrical pressure tanks, known colloquially as the *Torpedo Room*. This equipment is

part of the system that uses compressed air to pump sewerage up into the higher-level sewers that were designed by Joseph Bazalgette in the mid nineteenth century. Originally, when the Palace was first built, effluent was sometimes discharged straight into the Thames.

The roof above is vaulted brick and above that is the bottom of the air shaft. Crossing an access walkway over the plant we see the brick arch of the weight shaft and a large pile of sandbags on the floor. Above us, the shaft soars up out of view; with the help of a very powerful torch the weights can just be discerned way above. It was into this place some 150 years ago that the Great

Bell was carefully manhandled and then hauled up to the belfry over 200 feet above.

Our final visit is to look at the north side base of the tower. Here we see the stone that com-memorates the 100th anniversary of the clock, bells, and completion of the Clock Tower. Our exhaustive virtual tour is finished.

Finally, looking upwards, we see the Clock Tower soaring above us, that great and impres-sive icon of Great Britain.

CHAPTER THREE

THE PALACE OF WESTMINSTER TO 1834

THE BEGINNING

Like most ancient buildings, the Houses of Parliament—the Palace of Westminster—has evolved as the result of a long and sometimes complex series of events. This brief introduction is intended to put the building and its history into context, so that the reader can better understand how and why the Great Clock and the bells as we know them today came about.

THE ABBEY AND ITS LOCATION

Around 960, in the reign of King Edgar, Dunstan, Bishop of London, founded a Benedictine monastery on an area known as Thorney Island situated alongside the Thames and to the west of the City of London. There had previously been a church and monastery on the site, built early in the seventh century, but owing to attacks by invading Danes, the church

FIG 3.1 An impression of the first Abbey, from a Victorian magic lantern slide.

was abandoned until Dunstan revived the site. As the abbey evolved, it became known as West Minster, and hence Westminster, to distinguish it from the minster or cathedral of St Paul in the City of London.

The site for a monastery or abbey would usually have been chosen carefully by its founders, taking full advantage of nearby water and food supplies, and accessibility to local highways. Westminster, having evolved from small beginnings, did not enjoy such advantages. In the early Middle Ages, the locality was very different: indeed it was termed by historians a *loco terribili*; a terrible place, probably because of the combination of the extensive mud flats, the danger of flooding due to high tides or heavy rains, and the possibility of such diseases as malaria. Thorney eyot or island was a gravel bank bounded on one side by the Thames and on the other two sides by tributaries of the river Tyburn, which diverged somewhere around the present day site of St James's Park. As an island, its ground was a little higher than the mud flats that probably accounted for much of the surrounding banks of the Thames. The name 'Thorney' implies that it was once home to trees like blackthorn or hawthorn, or to common brambles.

It is thought that there has been a river crossing at Westminster since Roman times; this would initially have been a ford. Rivers today are most often seen as barriers; in the past rivers were the motorways of their time. Early maps show jetties at Westminster for landing or for ferries, but it was not until 1750 that a Westminster Bridge was completed to provide a permanent river crossing.

The first church was later replaced by the present Abbey; rebuilding was begun by Henry III in 1245 and today the abbey is one of the most important Gothic buildings in the country. Apart from being a place for personal worship, the Abbey is home for such national events as State services, coronations, and the weddings and funerals of the Royal Family.

A ROYAL PALACE

King Canute may have had a palace near the Abbey, but it was Edward the Confessor who appropriated the Thorney Island site that spread over some 5 acres. He established his Royal Palace here some time around 1052, and the site remains a Royal Palace to this day. Edward's palace may have been between the Abbey and the river, rather than where the present palace is. William the Conqueror set up his royal court here after his coronation in the Abbey, and the court sat here until the time of Henry VIII. After a fire in 1512, Henry VIII left Westminster, setting up court in York Place, which became known as Whitehall, in the 1520s. This remained a Royal residence until 1698; today Whitehall houses many Government offices.

WESTMINSTER HALL

William the Conqueror's son, William Rufus, built Westminster Hall, now the oldest surviving part of the ancient palace. The Hall was built between 1097 and 1099 and probably had two rows of pillars to support the roof. Even today it is a huge structure, so when built it must have been the wonder of the day. Successive Kings held their courts at Westminster, having been crowned in the Abbey. In 1399, Richard II completed improvements on the

FIG 3.2 The North door of Westminster Hall, circa 1800. Note the buildings that had been added onto the right-hand side.

FIG 3.3 The inside of Westminster Hall, showing the magnificent hammer-beam roof. Etching and aquatint by Rowlandson and Pugin produced for the 'Microcosm of London' series, originally published 1808–10. This figure is reproduced in colour in the colour plate section.

Great Hall by building the magnificent hammer-beam roof.

The Great Hall of the Palace has been used for major events such as Coronation feasts and the lying in state of monarchs. The Royal Courts of Justice met here and the Hall saw major trials, such as that of Charles I. As the Palace grew, it gathered the trappings of government: an Exchequer, Treasury, and various courts of justice, along with all the court officials needed to administer the various functions. The Great Hall became known as Westminster Hall.

ST STEPHEN'S CHAPEL

King Stephen (1135–54) built the first chapel on the Westminster Palace site, dedicating it to his namesake, St Stephen. A new Royal Chapel of St Stephen was begun by Edward I and completed in 1348. When it was finished, Edward III founded the College of St Stephen there, including a cloister and living accommodation for the colleges' canons.

THE FIRST CLOCKS

Sir Ralph-de-Hengham, when Chief Justice of the Kings Bench, caused a court roll to be erased and the fine of a guilty person to be reduced. To make an example of him, Edward I imposed a fine of 800 marks. The money paid was used between 1288 and 1290 to erect a tower containing a clock and bell, which struck the hour to remind the judges who sat in the adjacent courts of Westminster Hall of the offence of one of their predecessors. The clock bell was known as 'Great Tom of Westminster'. This must have been the first public clock to adorn the Palace of Westminster, but its exact position is unknown.

From 1365 through to 1367, a clock tower was built at the Palace for Edward III (probably named St Stephen's Bell Tower). In 1368, three Flemish clockmakers were granted free passage for a year to carry out their trade in England. It is possible that these men made the clock for the Westminster tower. From 1 August 1369, John Nicole was paid 6 d. a day for the care of the clock. On the tower was the motto *Discite justitiam moniti (Remember to learn justice)*.

In 1428, Agnes, the widow of Geoffrey Dalavan, petitioned the King for 100 s. 10 d. in payment for maintaining the clock. Geoffrey was also The Keeper of the King's Artillery, so it seems that Geoffrey was an official who managed various activities, including the winding up of and repairs to the Palace clock. Detailed in the petition was repair work done by Thome (Thomas) Clockmaker, including his *rewarde for the sayd yere*, indicating that he had an annual agreement to care for the clock and probably wind it as well.

A petition of 1455 to the King records that the Dean and Canons of St Stephens Westminster were paying 6 d. a day for the keeping of the clock. This must have been the clock tower and it would have been the Dean's immediate responsibility for winding and maintenance of the clock.

TWO VIEWS OF WESTMINSTER *(From Original Etchings by Hollar, 1647.)*

FIG **3.4** Two engravings by Hollar from 1647 reproduced in a nineteenth-century book, *Old and New London*. The top one shows the Palace of Westminster from the Lambeth side of the river, the lower one shows New Place Yard and the fourteenth-century clock tower.

The Old Clock-House at Westminster.

FIG 3.5 Westminster Clock House. After an engraving by W. Hollar reproduced in *The Mirror or Literature, Amusement, and Instruction* of May 1828. The text records that the inscription under the dial reads *Discite justitiam moniti.*

An engraving of 1647 by Hollar shows the clock tower. It was positioned opposite the north end of Westminster Hall and on the side of New Palace Yard. In between the Hall entrance and the clock tower was a fountain surmounted by a cupola that would have been around 40 feet high. When New Palace Yard was excavated in 1972–4 to build an underground car park for MPs, the base of the fountain was discovered. It was also determined that the ground of New Palace Yard was once mud flats that had been reclaimed as the river level fell and as cobbles were laid on successive layers of accumulated debris.

The large bell in the clock tower was originally known as 'Edward of Westminster' (possibly after Edward the Confessor), but later came known as 'Great Tom' or 'Westminster Tom'. The inscription, if accurately transcribed on later castings, suggests that the original bell was cast as a clock bell:

King Edward III made and named me
So that by the grace of St Edward the hours may be marked.

In 1698 the clock tower was in a bad state of decay and it was given by the King with some of the clock bells to St Margaret's church of Westminster, while the tower itself was demolished in 1707.

Great Tom the bell was sold to the cathedral of St Paul's; it weighed 4 tons 2 cwt, a truly massive bell for its time. However, whilst being transported to the City, Great Tom fell off its wagon and was cracked. It was recast several times by various founders until Richard Phelps of the Whitechapel foundry finally cast a satisfactory bell in 1716. The inscription on the bell records that the bell was brought from the '*Ruins of Westminster*'.

As near as can be ascertained, the old clock tower's original position would have been on the south pavement of Bridge Road, exactly opposite the north end of Westminster Hall. The present clock tower is only 100 yards from where its predecessor once stood.

HOUSES OF PARLIAMENT

Monarchs summoned a Parliament from time to time; a meeting of Barons, nobles, sheriffs, etc., to consider affairs of State and, more often, the raising of taxes for the King. In 1264, King

FIG 3.6 A View of Westminster from St James's Park. From left to right: the Clock Tower, Westminster Hall, the Abbey, and St James's Palace. Published in 1809, the view is based on an engraving by W. Hollar of around 1660. Note the wonderful open aspect that this part of London enjoyed in 1660.

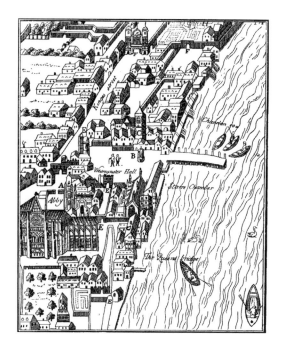

FIG 3.7 Aga's map of London, circa 1560. A, Abbey; B, Fountain; C, Clock Tower; D, Palace; Westminster Hall is named; Old Palace Yard is the area to the left of the Palace roof, marked D; New Palace Yard is the area around the fountain bounded by the river, the Hall, and the gatehouse.

Henry III was not popular with his nobility; a civil war developed and Simon de Montford called a 'parliament' in 1265 of two knights from each shire and two burgesses from each borough. Simon was later defeated by the King, but subsequent monarchs consulted more and more with their subjects on key issues of State. By the fourteenth century, two distinct bodies had evolved; the House of Commons and the House of Lords. Parliament met at the Palace of Westminster and has done so for the past 700 plus years, except for minor breaks.

The House of Lords had their own chamber in which to meet and debate but the House of Commons originally had no fixed meeting place to call their own. During Henry VIII's programme against the monasteries, when the Chantries Act of 1547 abolished all private chapels, the Royal Chapel of St Stephen within the Palace of Westminster was handed over to the House of Commons. Thus the House of Commons was permanently established in the old chapel.

A ROYAL RESIDENCE NO LONGER

Fires broke out at various times in the history of the Palace but in 1512, during the reign of Henry VIII, a serious fire damaged much of the Palace of Westminster, forcing the Royal Court to move out. The Westminster Palace was repaired but the Royal Family never returned to Westminster to live. Westminster Palace remained as a centre for the offices of government and for the sessions of Parliament.

After Cardinal Wolsey was deposed in 1529, Henry took over the Archbishop of York's Palace at York Place, a short distance down the river. There he built a new Palace, known as White Hall, after the pale stone used in its construction. Later, monarchs moved out of Whitehall and, like Westminster, the Whitehall area remained in State possession and today houses major government ministries, such as the Home Office, Foreign Office, Treasury, and Ministry of Defence. The only remaining parts of the old Palace are the wine cellars, the Banqueting Hall, and some river steps.

MAPS

Looking at maps of Westminster at various times, we see how the palace grew. Aga's map of around 1560 shows the Palace area alongside the Abbey. A view believed to date from around 1660 and reproduced in 1809 shows a view from St James's Park and depicts the Hall, Abbey, and Clock Tower. Anyone travelling in London now must be amazed at how open and rural Westminster was in the seventeenth century. By the late eighteenth century, buildings were starting to crowd in.

THE NINETEENTH CENTURY

As the years rolled on, the Palace was repaired, altered, and updated to meet the demands of the administration. By 1834, the Palace of Westminster centred round Westminster Hall. Numerous buildings had been erected on the west side of the Hall; these held the Exchequer, the Chancellor's office and Courts of Bail, Common Pleas, Chancery, and the Kings Bench. Opposite, on the East side, was the old St Stephen's chapel, which served as the House of Commons; next to this to the north on the river front was the Speaker's House. To the south were more buildings, including the House of Lords, libraries, the Royal Gallery, the Painted Chamber, and several coffee houses. New Palace Yard was the open area to the north of Westminster Hall.

The buildings of the Palace were mainly all connected together; many were made of wood and it was said that the roofing material used was often tarred sailcloth. Looking back with hindsight, the dark issue hanging over the Palace of Westminster was not '*Would there ever be a major fire?*' but '*When would there be a major fire?*'

On 16 October 1834, the inevitable happened, a fire broke out, and the Palace was almost completely destroyed.

CHAPTER FOUR

THE GREAT FIRE OF 1834

1826

In October 1826, Parliament passed a Bill that oddly enough initiated the building of the most famous clock in the world, the Great Clock of the Palace of Westminster. This bill made various reforms, one of which was the abolition of the tally-stick method of keeping accounts. The National Archives describes tally sticks as:

Used by accountants and by officials of the Exchequer who managed the revenue of the Crown. They were a physical proof of payments made into the Treasury. The Dialogue of

the Exchequer describes a tally as: 'the distance between the tip of the forefinger and the thumb when fully extended . . . The manner of cutting is as follows. At the top of the tally a cut is made, the thickness of the palm of the hand, to represent a thousand pounds; then a hundred pounds by a cut the breadth of a thumb; twenty pounds, the breadth of the little finger; a single pound, the width of a swollen barleycorn; a shilling rather narrower than a penny is marked by a single cut without removing any wood.' After the notches were made on the stick, the shaft was split lengthways into two pieces of unequal length, both pieces having the same notches. The longer piece (the stock or counterfoil) was given to the payer and the Exchequer officials retained the shorter piece (the foil). When the accounts were audited, the two pieces were fitted together to see if they would 'tally'.

Published by A. Beugo, Printseller, N° 38, Maiden Lane, Covent Garden Nov 12. 1810.

THE SPEAKER'S HOUSE, *From West Bridge*.

Built by M. Wyatt, in the year 1807.

FIG 4.1 The Speaker's House from Westminster Bridge, 1810. The gable end with perpendicular windows is St Stephen's Chapel, the upper part of which was used for the House of Commons. Published by H. Beugo, 38, Maiden Lane, Covent Garden.

FIG 4.2 Old Palace Yard, 1796. In the background is the south end of Westminster Hall. To the right is the entrance to the House of Lords where the fire began. Del. Miller. Published Colnaghi & Co., 132, Pall Mall.

OLD PALACE YARD.

1834

At 5.30, on the evening of Thursday, 16 October 1834, two workmen completed their day's work at the Houses of Parliament. They banked up a heating stove in the House of Lords with unwanted Exchequer tally sticks that they had been instructed to destroy by burning along with other old documents. The stove in the House of Lords was connected to iron flues, which passed through various rooms before venting into the atmosphere. Parliament was in recess, so the activity of burning tally sticks caused no interference with the normal business of the House.

At 6.35 that evening, a Mr Cottle finished his writing in a House of Lords' Committee Room. On leaving, he saw a great blaze in the Throne Room. Obviously, the fire had taken hold with great rapidity. Cottle raised the alarm. There was only just enough time for the Housekeeper and her staff to escape. By 7 p.m., the whole of the

House of Lords was ablaze. A stiff south-westerly breeze fanned the flames and within half an hour the fire had spread to the House of Commons, just as the roof of the House of Lords fell in. At 8 p.m., the whole of the House of Commons was burning fiercely. The wind increased to worsen the situation and by 11 p.m. the House of Commons roof fell in; the Library tower was to follow an hour later. The fire reached the Speaker's House, but that was only partially damaged; the Parliamentary Offices were saved. By midnight, it was obvious that little was to remain of the Houses of Parliament.

Fire engines were soon on the scene, but a low river and an ebb tide meant that water was in short supply. The police got there quickly as did the Coldstream, Fusilier, and Grenadier Guards. All activity was hampered by the crowds that gathered in a huge throng to watch the biggest fire London had even seen since the Great Fire of 1667. Confusion reigned: the fire brigades were all independent so lacked a common chain of

VIEW of the HOUSES of LORDS & COMMONS, destroyed by FIRE on the night of the 16ᵗʰ October 1834.

FIG 4.3 A chromolithograph, captioned 'View of the Houses of Lords & Commons, destroyed by Fire on the night of 16 October 1834'. Sketched from Serle's Boat Yard. This figure is reproduced in colour in the colour plate section.

command. However, Lords Munster, Adolphus, Fitzclarence, Melbourne, and Duncannon directed activities. Lord Duncannon was on the roof of the House of Commons with firemen until it was too risky to stay; he was the last to leave the roof and two minutes later it collapsed. The ancient Westminster Hall seemed to be in great danger of being burnt as well, so once it was obvious that Parliament was lost all activity was directed at keeping the flames away from the Great Hall. A large number of documents and books were rescued from the Houses of Parliament before the fire had spread everywhere. These were put under carpets and tarpaulins and then moved to Downing Street or into St Margaret's church for safe keeping. The Chancellor's Mace and the Mace of the Sergeant at Arms were saved. As the fire took some time to get to the Speaker's House, his library, a marble fireplace worth £200, his plate, and his wife's wardrobe and jewellery were all safely taken out.

Apart from local brigades, fire engines came from the Horse Barracks and Elliot's Brewery, and two parties of marines brought two engines up from Deptford Dockyard. In the early hours of the morning a steamer arrived from Rotherhithe towing a floating pump. This was soon in action, throwing water at the rate of a ton a minute into the midst of the conflagration.

At midnight, when the blaze was at its peak, the whole of London seemed to have turned out to watch. The best viewpoint was on Westminster Bridge. Turner's picture *The Burning of the Houses of Parliament* shows a massive wall of flame rising hundreds of feet into the air; above is a huge plume of fire, sparks, and smoke. Turner shows the bridge and south bank crammed with watchers and the whole scene illuminated with a deep orange glow.

The floating pump was highly effective and doused the flames. By morning light, there was just a mass of smouldering ruins with a few little fires still burning. The Royal Family visited the ruins and the clearing operation started. Fortunately, no one was killed, but some eight broken limbs were sustained, mostly due to traffic. The police had apprehended a multitude of petty thieves and pickpockets, all of whom were brought before a Court held the following day.

Before the fire, Mrs Wright, Housekeeper to the House of Lords, complained of the excessive heat in the chamber. She had a lad called David Reynolds, who was employed to answer the door, he was sent on several occasions to tell the workers to reduce the level of the fire; they denied ever having received the messages from Reynolds. Mrs Wright also fancied that she detected a slight smell of burning, but did nothing about it. Around four in the afternoon Mrs Wright had shown a party of visitors round the House of Lords, in one place the floor was so hot it could not be touched; there was a strong smell of burning and there was smoke in the chamber.

The Privy Council prepared a report on the fire, laying the blame on the excessive amount of wood that the workmen had put into the stoves. They were also very critical of Mrs Wright's failure to contact an appropriate authority when she smelt burning. No malicious intent was found in anyone. Mr Cooper of Hall & Cooper, the stove makers, said that he heard in Dudley, at 10 p.m. on the Thursday night, that the Houses of Parliament were burning. Since the information could not have travelled the 120 miles in a few hours, then the narrative of Mr Cooper implied that the fire must have been arson. A detailed investigation could not corroborate his account, which must have been caused by Cooper confusing times and dates; the news had certainly reached Birmingham the next morning.

The report on the causes of the fire read just like a modern-day horrific accident, where a chain of events, rather than a single episode, were responsible for the outcome. The Surveyor decided to burn the tallies in the stoves and not in the Palace Yard, as initially directed. The housekeeper, Mrs Wright, thought that she smelt burning but did not report it. The workmen were not properly supervised and put too many

FIG 4.4 Cloister Court and the remains of St Stephen's Chapel, from *The History of the Ancient Palace of Westminster*. E.W. Brayley & J. Britton, 1836.

tallies on the fire. *The Times* reported that heating flues had recently been repaired and some experiments had been carried out, and a cost-saving measure that reduced the number of attendants in the House of Lords, depriving the establishment of some of the individuals who had the care of the flues. Also, the flues had not been swept for a year.

Overall, the excessive burning of wooden tallies on stoves intended for coal, coupled with what was probably a faulty flue, were the likely causes of the fire. Wood next to a flue had most likely been smouldering for hours before it exploded in flames. This would have explained the great rapidity with which the fire took hold. One of the workmen, named as James Cross, became the scapegoat and was dismissed, but no one else was reprimanded.

An onlooker at the latter part of the fire was Charles Barry, an emerging architect who, on returning from Brighton, saw the red glow in the sky and hastened to watch the scene of destruction. The same day, Edmund Beckett Denison, an 18-year old lad from Yorkshire recorded in his notebook: *16 Oct 1834, the day the Houses of Parliament burnt down. Went up to Cambridge.* Both were to meet as bitter enemies in the future and both were to make significant contributions. Barry became the architect of the new Houses of Parliament and Edmund Beckett Denison became a lawyer and the designer of the Great Clock and the bell *Big Ben*.

CHAPTER FIVE

A NEW PALACE: 1835 ONWARDS

1834: AFTER THE FIRE

Hundreds of years of history, in the build-ings of monarchs and Parliament, were swept away in just one night by the fire. Now what should be done? Westminster Hall escaped, as did various administration buildings, but all the key old buildings were gutted. Temporary roofs were fitted on the shells, sufficient to provide a working place for the House of Commons and the House of Lords.

But the damage had been done. It was decided that all the remnants of the old Palace be demol-ished, with the exception of Westminster Hall, and that a new Houses of Parliament was to be built.

1835: THE NEW BUILDING

Early in the year, a Royal Commission was set up to oversee the design of the building of the new Houses of Parliament. A competition was launched: the new design had to be in the Elizabethan or Gothic style. Plans were to be submitted by 1 November. Some 97 entries were received; just six months were allowed for the designs to be prepared after the competition had started.

Charles Barry

Charles Barry was born in 1795. His father was a stationer and Charles was able to travel and study architecture, which he took up as his preferred

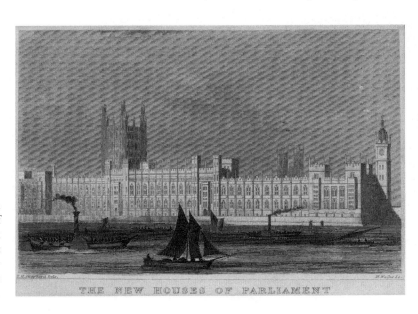

FIG 5.1 Barry's winning plan for the New Houses of Parliament. The Clock Tower was not a requirement of the competition but Barry added one after he was declared the winner.

THE NEW HOUSES OF PARLIAMENT

profession. From humble beginnings, he went on to design a host of prestigious buildings. In 1826, he designed St Peter's church in Brighton, which was acclaimed as a fine example of Gothic revival.

With the competition declared for the new Houses of Parliament, Barry engaged Pugin and together they produced a design for the New Houses of Parliament. Barry was the mastermind who had the vision for the Palace design, layout, and construction. Pugin was obsessed with the Gothic style and he contributed to every aspect of decoration in Barry's plans.

Barry's design was the winner and, after years of work, he received a knighthood in 1852. Barry died suddenly in 1860; just a year after the Clock Tower was finished and with much work incomplete at Westminster. Barry had four sons, three of whom followed him in architecture. Edward completed his father's masterpiece of the Houses of Parliament. Charles was buried in the nave of Westminster Abbey in honour of his contribution to the creation of the New Palace of Westminster. His gravestone is embellished with a brass showing the layout of the Houses of Parliament.

Augustus Pugin

Augustus Welby Northmore Pugin was born in 1812, the son of a French artist. Augustus inherited his father's talent and became intensely interested in architecture, particularly in the Gothic style. He went into architecture and was soon designing or renovating churches, the Gothic style being used exclusively. Pugin converted to Roman Catholicism and designed many Catholic church buildings.

Barry employed Pugin to help with his plans for the Houses of Parliament. Pugin was a genius

SIR CHARLES BARRY

FIG 5.2 Charles Barry, the successful architect of the New Houses of Parliament, from Cassell's *Old and New London.*

who applied the Gothic style to every aspect of decoration in the Palace from the floor tiles to carpets, through wall hangings, stained glass windows, and even the iron finial on the top of the Clock Tower. Barry was the mastermind who designed the Palace, its layout and function, but it was Pugin who applied all the elaborate Gothic ornamentation.

It is said that genius and insanity are never far apart, and this was indeed the case with Pugin; he had episodes of illness alternating with periods of intense activity. He designed the Clock Tower at Westminster and was only 40 years old when he died in 1852 from a bout of insanity, which was attributed to a possible brain tumour.

FIG 5.3 Pugin, as depicted in his obituary in the *Illustrated London News*.

Barry did not acknowledge Pugin's input at Westminster and the surviving families never had good relationships.

1836

In March, Committees were formed in the House of Commons and the House of Lords to report on plans for the proposed new Houses of Parliament. A decision was made in favour of number 64, the design by Charles Barry.

A display of the unsuccessful designs submitted was proposed and, when finally agreed, this was put on in the National Gallery. The display was visited first by the King and then by Members of Parliament and Peers, after which it was thrown open to the public on 24 March. One design incorporated a house for the resident architect, another was submitted by a patient in Bedlam, who was mad when the wind was in the south but at other times designed architecture! Barry's plans were not displayed, much to the disappointment of the public. Lord Duncannon and the Commissioners kept them in their office to progress the work on the New Palace.

BUILDING THE PALACE OF WESTMINSTER

No sooner than Barry was declared winner and appointed architect, many of his profession turned on him in a fit of jealousy. There were attempts to have the competition re-opened, but Barry had won, and he won by the merit of his design and not by his position or by having friends in high places.

Creating some solid foundations took considerably longer than expected: a coffer dam was built along the river front and, once the water was pumped out, thousands of tons of concrete were poured in. Once Mrs Barry had laid the foundation stone on 27 April 1840, the real work of building could commence. Grissell and Peto won the building contract; Peto was later to get a knighthood and a large fortune for his building endeavours.

Barry was never far from conflict; first, regarding his fee, which the Government was intent on reducing to a minimum. Then he was inflicted by more managers than anyone could reasonably cope with. After the First Commissioner for Works, there were the House of Commons, with their demands, the House of Lords, with their agenda, the Treasury, the Lord High Chancellor, who represented the Crown, and many others.

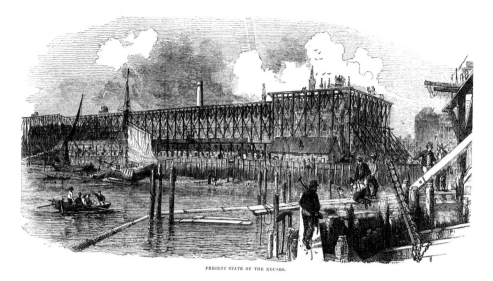

PRESENT STATE OF THE HOUSES.

FIG 5.4 A view from the river, in 1842. The coffer dam to keep the water out is clearly seen and the building is shrouded with scaffolding. It looks as though the walls are at about two-thirds of their final height; this is probably a speculative situation. *Illustrated London News.*

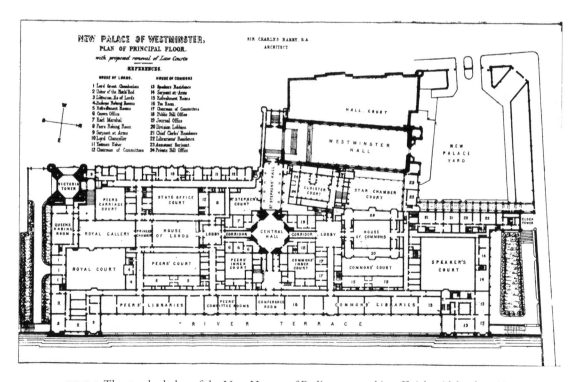

FIG 5.5 The standard plan of the New Houses of Parliament used in official guidebooks, 1882.

One significant player in the construction of the Palace was a Dr David Reid, who had convinced the Government that he was an expert in heating and ventilation. Reid was engaged to incorporate his inventions into the Palace. Barry was never consulted on the matter, and Reid's demands on Barry were significant, requiring extensive arrangements of ducts and flues to be included into the building. It was said that one-third of the Palace's volume was to be taken up by Reid's needs. One major requirement was that the Clock Tower had to contain a major air shaft that ran from the first storey all the way up to the belfry. This measured around 8 feet by 16 feet; initially the idea was to draw fresh air in at the top of the shaft and then distribute it through the Palace. Although Reid was dismissed in 1852, work on the Clock Tower had started and the air shaft was built. The whole of the ventilation system was a problem; by the time the clock came to be installed, Barry had altered the ventilation scheme so that the air shaft was being used as a chimney. It seems that it was used as such until around 1914, when electric fans were probably installed.

Goldsworthy Gurney, whose name can be found on massive cast-iron stoves sometimes seen in English cathedrals, replaced Reid as the heating engineer. At one stage, he had a vent from the sewer connected to the Clock Tower air shaft, with the idea that a flare could be burnt on the top of the tower and thus draw air upwards. It did not work, so a fire was provided at the bottom of the tower to create an updraught. A gas main leaking into the sewer caused a small explosion. Had the gas got into the airshaft, then the Clock Tower might have been demolished by a massive explosion that would have competed with the fire of 1834 for spectacle.

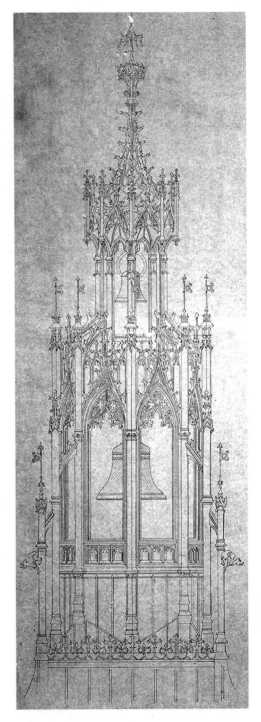

FIG 5.6 An early drawing of the bell cupola, probably by Pugin. Courtesy of the National Archives.

44

THE CLOCK TOWER

When it came to designing the clock tower, Barry had problems; he could not get the design to look correct. Early engravings of the proposed Palace show a small clock tower of insignificant proportions with a cupola on the top. Pugin was engaged by Barry to take on the massive task of designing the Clock Tower.

One of Pugin's early commissions was the remodelling of Scarisbrick Hall in Lancashire. The result was that the Hall is almost a Houses of Parliament in miniature, profuse in the Gothic decoration so beloved of the architect. The oriel windows, roof, decorative ironwork, and statuary all pre-empt the decorations at Westminster. Pugin started the work at Scarisbrick in 1837 and by 1839 had completed the clock tower. Sadly, Pugin's, clock tower was replaced by his son at the request of the owner. Today, this new tower dominates the building and is quite out of proportion with the rest of the Hall. We are

fortunate that in the Hall is a piece of carving showing the Hall as it was originally, including the first clock tower. The lantern top, the tiny dormer vents, and the protruding dials are all unmistakeably the same as the top of the clock tower at Westminster. There can be no doubt that when Pugin designed the clock tower at Westminster, he had Scarisbrick in mind.

In her book *God's Architect*, Rosemary Hill quotes a letter written by Pugin in February 1852 to his friend, Hardman, who was a Birmingham manufacturer of ecclesiastical ornaments. '*I never worked so hard in my life for Mr Barry for tomorrow I render all my designs for finishing his bell tower & it is beautiful & I am the whole machinery of the clock.*' At this date the clock tower had risen to beyond the roof level of the House of Commons. So it looks as though the design of the upper part was completed just when Pugin was slipping into madness and some seven months before he died. Sadly, Pugin never saw the completed tower. Augustus would have been much comforted on

FIG 5.7 A conjectural reconstruction of Scarisbrick Hall as it probably was in 1850, with the Westminster-type clock tower.

his final sick bed if he was able to know that his masterpiece was to become one of the most famous and well-known buildings in the world and the iconic representation of Great Britain.

Barry made no recognition of Pugin's input to the design of the Palace. After Pugin died, his son Edward went to Barry with Augustus's letters to claim a place in history for his father. Barry asked for the loan of the letters and they were never seen again.

ELEGANT PROPORTIONS

The Westminster Clock Tower is undoubtedly a masterpiece. Viewed from Bridge Street, the tower elegantly soars up from a slightly enlarged base; its slenderness enhanced by the tall verticals and long narrow windows. To make the top look more pronounced, the dials are jettied out from the main walls. Oddly enough, the dials only occupy half the total width of the tower making them look somewhat small in a line drawing. In reality they look exactly right; the white dials against the honey-coloured stone viewed from the ground appear just perfect. Above the dials, openings allow the sound out from the belfry. Higher up, the roof tapers into the lantern and above that the spire progresses upwards terminating in a gilt crown and finial.

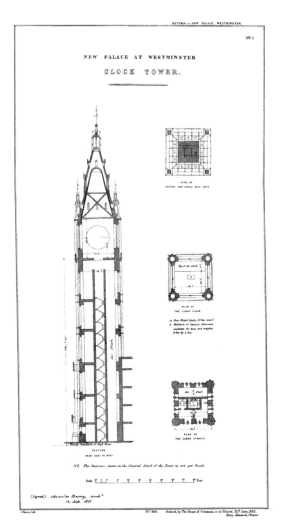

FIG 5.8 Barry's plans of the Clock Tower included in papers relating to the Great Clock, dated 12 September 1851; published by the House of Commons in 1852.

BUILDING THE CLOCK TOWER

There is no formal foundation stone for the Clock Tower. In the Parliamentary Archives there is a silver trowel that bears the inscription: *The First Stone of the Clock Tower of the New Houses of Parliament was laid by Emily, second daughter of*

Henry Kelsall, Esquire of Rochdale. 28 September 1845. It also names Charles Barry as Architect and Thomas Grissell and Samuel Morton Peto as Builders. Emily was the sister of Peto's daughter-in-law. It appears that Henry Kelsall was a mill owner in Rochdale.

A new method of building was employed on the Clock Tower: it was built from the inside up. In construction, the tower is brick-built and

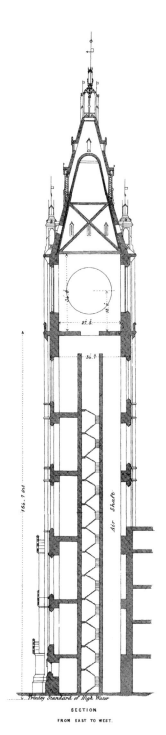

FIG 5.9 The vertical section of the Clock Tower in more detail.

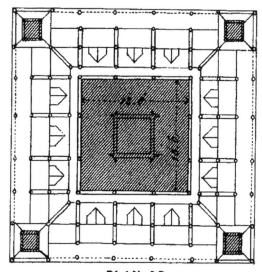

PLAN OF
CENTRAL AND ANGLE BELL COTS

FIG 5.10 A detail of the plan at the belfry.

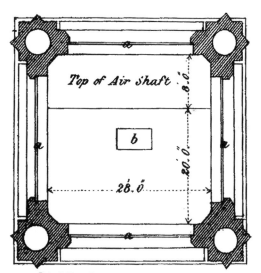

PLAN OF THE CLOCK FLOOR.

a. Four Metal Dials, 20 feet diam.
b. Wellhole of Central Staircase,
available for lines and weights,
6 feet by 4 feet.

FIG 5.11 A detail of the plan at the Clock Room.

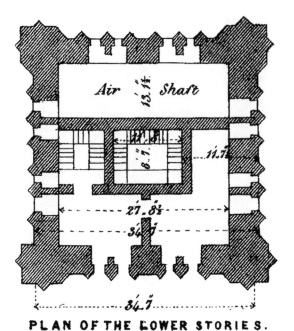

PLAN OF THE LOWER STORIES.

FIG 5.12 A detail of the plan at the lower floors.

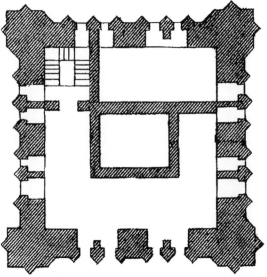

FIG 5.13 A detail of the plan at low level, modified to the present arrangement of weight shaft and stairs.

faced on the outside with Anston stone. Inside, the brick walls were covered with rendering and plaster. Wrought iron cramps were used to secure the stone facing to the brickwork. An internal scaffold system was jacked up as each course of stone and brick was laid. The spire is made of wrought iron and roofed with cast-iron tiles.

Barry was asked by the Commissioners in 1848 what savings could be made by delaying the building of the Victoria and Clock Towers, a classic case of a department embarrassed by overspend. Barry had to reply that no savings would be made; indeed a cost would be incurred, since stone had been ordered, scaffolding was hired along with two steam engines; and lifting equipment and cordage would seriously suffer from lack of use. Also, the engineer and his men were now familiar with their task and contented; these would have to be laid off. The work on the heating and ventilation would be delayed.

It seems the building of the tower made slow and erratic progress. From engravings of the time, we can track the growing of the tower but artistic license in the illustrations has to be taken into account. Some pictures are certainly imaginative, but overall we get the picture of progress from the State Opening of Parliament in 1852 through to the finished tower in 1860.

THE PALACE COMPLETED

As with any large building under construction, the Palace became usable in various stages. In 1847, The House of Lords sat in their new chamber for the first time and it had its official opening, Pugin was not invited. He was denied the wonderful spectacle of the Peers in their robes in the ornate gold and red setting that he had designed. The House of Commons moved into their new chamber in 1850, and promptly

FIG **5.14** The State Opening of Parliament on 3 February 1852. Note that the clock tower is almost up to the level of where the dials jetty out. From *The Illustrated London News.*

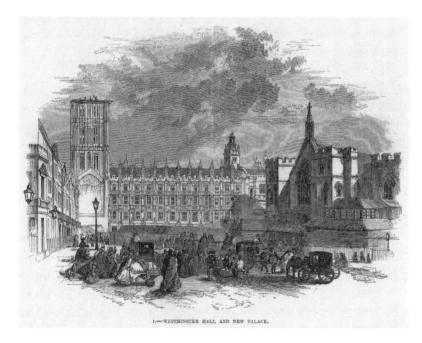

1.—WESTMINSTER HALL AND NEW PALACE.

FIG **5.15** A view of the Clock Tower from New Palace Yard, circa 1852. The House of Commons has been completed but the old houses on the west side of the yard still remain. The lack of external scaffolding shows that the Clock Tower was built from the inside.

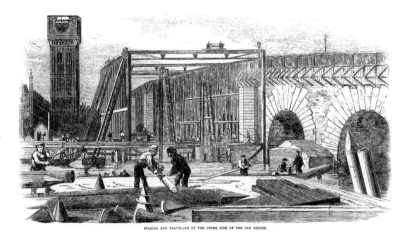

FIG **5.16** A view of Westminster Bridge, 3 February 1855. The clock tower is about three-quarters of the way up the dial openings. From *The Illustrated London News.*

STAGING AND TRAVELLER ON THE UPPER SIDE OF THE OLD BRIDGE.

FIG **5.17** Another view of Westminster Bridge, 3 February 1855. A roof to protect the works can be seen above the tower. From *The Illustrated London News.*

NEW WESTMINSTER BRIDGE—GENERAL VIEW.—THOMAS PAGE, ENGINEER.

moved out again complaining that the roof was too high and the acoustics were wrong. It was to be two years before Barry had made acceptable alterations and the MPs could move in. On 11 November 1852, Queen Victoria performed the official opening of the New Houses of Parliament at the State Opening.

In 1857, Barry proposed a massive extension to the Palace with offices enclosing the New and Old Palace Yards. There was to be an impressive entry arch in the north-west corner and Dent quoted for a clock for the portal. However the extension was never built and it was not until the twentieth century that Portcullis House was built to provide additional office space for MPs.

Barry died suddenly in 1860, of a heart attack, and his son Edward carried on as architect,

FIG 5.18 The Clock Tower and Speaker's House from
the Thames. Another interpretation of what the final
Palace would look like. From *The Illustrated
London News*, 11 April 1857.

Government to wind up matters that had been
allowed to come to a dead standstill.

THE COMMISSIONERS

Key players in the whole story of the clock and
the building of the Houses of Parliament were
the various Commissioners for Works. Between
1832 and 1850, the Government department that
was involved in building the New Palace was
known as The Commissioners of Woods,
Forests, Land Revenues, Works, and Buildings.
In 1851, it divided and the department dealing
with buildings, etc., took the name of The
Commissioners of Works and Public Buildings,
in short the Office of Works, a name which it
retained until 1940.

The First Commissioner was a Government
official. He ran the department and was assisted
by two co-Commissioners who were permanent officials. Since much of the correspondence concerning the Palace, Great Clock, and
Bells was conducted through the Office of the
Commissioner of Works it is useful to know
who were the First Commissioners during the
period concerned. Assistant Commissioners
did not change with each Government; Alexander Milne and the Hon. Charles Alexander
Gore were assistants from 1841 to 1851.

First Commissioners changed as the government administration changed when elections
were lost or won. In the nineteenth century,
there were two main political parties—the Tories
and the Whigs. The Tories gained their support
from Royalty and the aristocracy and the Whigs
from free thinkers and industrialists. Although
the nineteenth-century policies are not the same
as those held today, the Tories migrated into
today's Conservative party and the Whigs into

directing the completion of the work. It was not
until 1870 that the Palace of Westminster was
essentially completed. Edward Barry was to all
intents and purposes dismissed and the Palace
building was taken over by the Office of
Works.

On 15 July 1847, a waggish letter in *The Times*
noted that £3,300 had been designated for a new
clock for the Parliament building; the writer
hoped it would have an alarm to wake up
sleeping Ministers. He continued to wonder
if the clock would need winding up pretty
often, as this would serve as a reminder to the

FIG 5.19 A view of the completed Clock Tower from Victoria Tower. From *The Illustrated London News*, 28 January 1860.

FIG 5.20 An albumen print of the Clock Tower, taken circa 1880, from Parliament Square looking across New Palace Yard. The stonework is in a clean condition, an indication that much of the pollution took place in the twentieth century.

the Liberals. Over the history of the clock from inception to completion there were eight different first Commissioners; three Tory and five Whig. The key Commissioners were: Lord Seymour, who appointed Sir George Airy as a referee; Sir Benjamin Hall, who gave his name to the Great Bell; and Lord John Manners, who kick-started the project to procure the bells back into action after it had ground to a halt.

Here is a resume of First Commissioners and what happened concerning the clock, bells, and tower during their term of office.

The Earl of Lincoln, Henry Pelham-Clinton	1841–6 (Tory)	During his time Barry first made enquiries to Vulliamy about a new clock.
2nd Viscount Canning	1846 (Tory)	Viscount Morpeth; George Howard 1846–50 (Whig)
Lord Seymour	1850–2 (Whig)	It was during Seymour's time that Sir George Airy was appointed a Referee for the Great Clock
Lord John Manners	1852 (Tory)	In Manner's short term of office, the order for the Great Clock was placed with E.J. Dent.
Sir William Molesworth	1853–5 (Whig)	Molesworth was ineffective in getting the bells cast and dismissed Denison as a referee for the bells. His term of office was terminated by his sudden death.
Sir Benjamin Hall	1855–8 (Whig)	During Hall's time, the first Great Bell was cast and was named after Sir Benjamin. Later, Hall became Lord Llanover.
Lord John Manners	1858–9 (Tory)	Manners oversaw the recasting of the Great Bell and the final installation of the bells.
Henry Fitzroy	1859–60 (Whig)	Fitzroy saw the Great Clock installed and set going and the cracking of the second Big Ben.
William Francis Cowper	1860–6 (Whig)	It was during Cowper's office that the Great Bell was finally turned and set striking again.

CHAPTER SIX

PLANS, SPECIFICATIONS, AND TENDERS FOR THE GREAT CLOCK: 1844 TO 1852

THE REQUIREMENT FOR A GREAT CLOCK

It was in 1844 that the first move was made to secure a Great Clock for the new Palace of Westminster. To appreciate just what a significant advance in technology the making of the Great Clock demanded, it is worthwhile first to review the production of public (turret) clocks towards the latter part of the second quarter of the nineteenth century. In construction, the size of the dials, the accuracy of timekeeping and the weights of bells to be sounded, the Great Clock had to exceed commonly available clocks by a very wide margin.

TURRET CLOCKS

A turret clock is a public clock that shows the time on an external dial. A turret clock, say in a church, normally drove an outside dial that might be six feet, or even up to ten feet in diameter; sizes above this were quite unusual. Hours were sounded on a bell; a large bell of that time would be perhaps 30 cwt. If the clock struck the quarter hours, this was normally two notes, known as ting-tang, struck on two bells, one

ting-tang being struck at quarter past, two at half past, and so on, until it struck four ting-tangs before the hour was struck.

A turret clock of the early eighteenth century consisted of a wrought iron frame like a birdcage containing two trains of gears, the wheels probably being of iron. Winding was needed every day or every other day, depending on the drop available for the weights. By the 1840s, the design of English turret clocks had changed very little from that of a century earlier. Cast iron replaced the forged wrought-iron frame and brass wheels were employed in place of iron wheels. Often, an extra wheel had been added in each train that allowed the clock to run for eight days between windings; a much more convenient going period. The specification for the Great Clock was to demand a very different type of construction from what was commonly supplied.

Timekeeping

Accurate timekeeping was of no great importance in the eighteenth century, where a clock provided the time to a local community. In the nineteenth century mail coaches started to run on published timetables, so the differences in public clocks from one town to another began to be an issue. By the middle of the century, the

FIG 6.1 A late seventeenth-century clock. Movements similar to this continued to be made for another 100 years.

FIG 6.2 A four-posted movement of around 1830, probably made by Hale of Bristol.

boom in railways meant that accurate time keeping was essential for those who wanted to catch a train.

The timekeeping of a turret clock was variable: for steady weather a clock might keep good time for weeks, but with weather changes it might easily gain or lose a few minutes in a week. There was no real understanding by turret clockmakers of all the issues that affected timekeeping. There were three major causes:

temperature, circular error, and escapement error. Together, these errors could combine to give a variation in timekeeping of several minutes a week and this performance was not acceptable to a mid-nineteenth-century industrial society.

Changes in temperature caused the pendulum rod to change length and thus the timekeeping was altered; the difference between timekeeping in summer and winter could be 15 seconds a day when an iron pendulum rod was used. It was possible to build special pendulums that compensated for temperature changes, but these were very rarely used in turret clocks. Makers tended to use wooden rods for pendulums; these gave a better performance than iron but were affected by moisture unless they were well varnished.

Weather effects on the exposed dials included wind blowing onto the hands and ice and snow changing their weight and balance. Such effects would change the amount of power transferred to the escapement, either adding or subtracting depending on many factors. Changing power at the escapement meant that the arc of the pendulum swing changed. The mathematical formula for a simple pendulum has been derived assuming that a pendulum swings through a tiny arc. In real life, a turret clock pendulum will swing two or three degrees to each side of the vertical. In this case, the formula that defines the time of swing has to include the pendulum arc. Circular error is the term that clockmakers use to define the difference in timekeeping between a mathematical simple pendulum and a real pendulum swinging through an arc of several degrees. Providing that the pendulum arc does not change there is no problem, but if the arc changes due to variations in power caused by varying weather on the dials, then the circular error changes. In real life, this may account for something like ten seconds a day.

Changes in air pressure will affect the time-keeping of a pendulum; air has a weight and the pendulum bob essentially *floats* in air. Air density changes with temperature and humidity so this too has an effect. Fortunately the error is quite small, depending on the situation: it has been estimated to be in the region of 2 to 18 seconds a year.

Finally an escapement, as it impulses the pendulum, causes a variation from the ideal timekeeping of the pendulum. Escapement error depends on pendulum arc, so changes in arc lead to variations in timekeeping. Each clock has its own peculiar escapement error and this could cause errors ranging from a small variation to tens of seconds a day.

All of these errors add up to give poor time-keeping. A *good* clock might keep time to half a minute a week, a *poor* one might vary by several minutes. If *good* was 30 seconds a week, then the Westminster clock had to be *excellent*: it was to required to be accurate to just one second a day.

Dials

In early times public clock dials were mostly made of stone or wood but by the eighteenth century a circular sheet of copper was commonly used. This was beaten out to a convex shape and a moulding formed round the edge to provide rigidity. After painting, the minute marks and chapters were put on in gold leaf.

A clockmaker of London, J.P. Paine, who worked in Bloomsbury in the first half of the nineteenth century, had invented an illuminated dial around 1827. It had a frame glazed with coloured glass and was lit from behind by oil or gas light but it was very slow in gathering popularity. James Harrison of Hull made skeleton dials of cast iron in the early 1830s; these

he fixed to towers and the masonry could be seen through them. After 'Big Ben' such skeleton dials became common and were glazed with opal glass and illuminated from behind by gas.

The largest dials in England for the period were at St Paul's cathedral, in London, being two dials of 17 ft diameter, 120 feet above the ground. The largest dials for a church were Holy Trinity church in Hull, which had four 13-foot diameter dials that were installed by James Harrison of Hull when he modified the existing clock in 1840. The Great Clock was to exceed this; dials were to be over 22 feet in diameter and illuminated from behind.

FIG 6.3 Painted copper dials were common from 1700 through to the 1960s, when glass-reinforced plastic replaced copper as a lighter and cheaper material.

FIG **6.4** A skeleton dial of cast iron by James Harrison of Hull, North Cave church, 1851.

FIG **6.5** At 17 feet in diameter, the dials of St Paul's cathedral were the largest in the country.

Bells

Few churches boasted large bells, say 2 tons or more in weight, but some cathedrals often had very large bells. The clock at St Paul's cathedral was originally built by Langley Bradley in 1708 and later modified; it struck the hours on the bell 'Great Tom', which weighs just over 5 tons. An engraving in Benson's book *Time and Timetellers* shows the old clock, but no indication is given as to its size. However, the old escape wheel still exists and from this it may be deduced that the clock movement was something like 6 feet high and over 8 feet long. Today the old clock is lost, having being replaced by a new one in 1892.

REQUIREMENT

It was into this somewhat stagnant scene of turret clockmaking that the requirement for the Great Clock of Westminster was launched. The prerequisite was for four dials of around 25 feet diameter, and these needed to be some 200 feet above the ground. The clock was to ring the quarter hours, and the hour was to be sounded on a bell of 14 tons. Once a proper specification had been written, the accuracy required was that the clock to be accurate to within a second, the exact instant of the hour being the first blow on the bell.

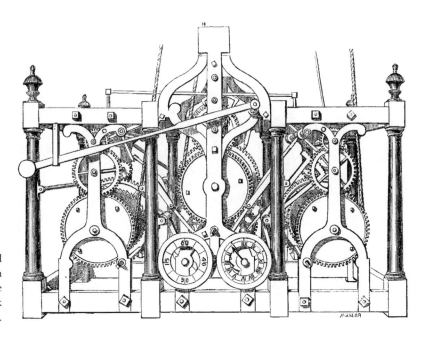

FIG 6.6 St Paul's cathedral clock was about 6 feet high and, in the 1840s, was the most powerful clock in England.

TURRET CLOCKMAKERS

It is worth considering the principal makers of turret clocks in operation around 1840–50. This helps to explain why E.J. Dent was finally chosen as the maker and why Vulliamy was approached by Barry.

Benjamin Lewis Vulliamy

The Vulliamy clockmaking dynasty came from Switzerland and started with Francois Justin Vulliamy, who was always known as Justin Vulliamy. It is believed that Justin trained as a watchmaker in Switzerland, spending some of his time in Paris. Justin emigrated to England, arriving in the 1730s and was soon working with Benjamin Grey, a famous watchmaker. Justin married Mary, Grey's daughter, and Grey and Vulliamy worked together in partnership producing clocks and watches. Justin was patronized by the King.

Benjamin was born to Justin and Mary in 1747 and took to the family trade, making watches and clocks. He was a particularly excellent craftsman and was appointed Clockmaker to the

FIG 6.7 A wax portrait of Benjamin Lewis Vulliamy, Clockmaker to the Queen. This figure is reproduced in colour in the colour plate section.

King in 1771. Benjamin married Sarah de Gingins from Switzerland and their first son Benjamin Lewis was born in 1780.

Turret clocks by Justin and Benjamin were made by John Thwaites, a well-known London maker. It was common practice in the clockmaking trade to buy in clocks from specialists, but to have the supplier's name added as though they were the maker.

Benjamin Lewis Vulliamy was to be the most famous of the clockmaking family. He excelled in everything he did and was a commercial success. As a shrewd businessman, he supplied not only clocks to Royalty and the aristocracy, but also furnishing items, such as fireplaces and candelabra. After the death of his father, Benjamin, in 1811, Benjamin Lewis inherited the business and followed his father's footsteps by being appointed Clockmaker to the King.

Much of the clock and watchmaking trade in that period was carried on by outworkers. Benjamin Lewis was able to manage them since he was well able to do all the jobs himself and he thus demanded and obtained excellence.

Benjamin Lewis was not only Clockmaker to Queen Victoria; he also held various posts in the Worshipful Company of Clockmakers and was five times its Master. In addition, he was a Fellow of the Geographic, Astronomical, and Zoological societies, as well as an associate member of the Institute of Civil Engineers.

Benjamin Lewis designed his own turret clocks; he had the castings made by the Bramah brothers and the wheel cutting, clock making, and installation done by William Vale and his son George Frederick. In 1828, Benjamin Lewis wrote a short book, *Some Considerations on Public Clocks*, in which he outlined a new method of construction of turret clocks. Traditionally, turret

FIG **6.8** West Norwood church clock by B.L. Vulliamy, 1827. This had a 'flatbed' design, which probably originated in France.

clocks had been built into a cage-like iron frame. Benjamin Lewis introduced a box-shaped base of cast-iron, onto which the supporting bars and wheels were added. Assembly and maintenance was thus made easier. Despite this invention, which Benjamin Lewis had almost certainly seen in France, he made only a handful of turret clocks with this plan, the first of which was installed at West Norwood church in London. This could be described as the first flatbed turret clock in England. In all, Benjamin Lewis produced about 100 turret clocks, some of which went abroad to the colonies.

Benjamin Lewis Vulliamy's involvement with the Great Clock is significant, although he did not get the contract to make the clock. However, it seems we can thank him for precipitating some of the publications of official correspondence that give us such a good insight into the Clock's history.

Thwaites and Reed

In the 1840s, Thwaites and Reed were the largest makers of turret clocks in the Kingdom; they supplied many Government departments, such

FIG 6.9 A movement by John Thwaites: the company later became Thwaites and Reed. Although dated 1801, the design used by this maker was still very much the same in 1840s.

as the Board of Ordnance and the East India Company. Aynsworth Thwaites was the founder; he was apprenticed to Langley Bradley, who supplied a large turret clock to St Paul's cathedral. Aynsworth's most famous turret clock was supplied to the Horse Guards in London in 1768. John followed his father Aynsworth; the company becoming Thwaites and Reed in 1828. Around 1856, the company was bought by a Mr Buggins; this family was involved with the company until 1978.

Clocks by Thwaites and Reed were well made. Many of their clocks of the early nineteenth century are still in service and show little signs of wear on wheels or pinions. The company was traditional in its approach and probably would have proposed a massive birdcage movement had they been invited to tender. The clocks they were making around 1840 looked much like their mid-eighteenth-century movements, except that cast iron was used instead of wrought iron for the frames.

John Moore and Sons

At the end of the eighteenth century, two of Thwaites's apprentices, Handley and Moore, set up their own clockmaking business. Early Handley and Moore turret clocks are virtually indistinguishable from clocks by Thwaites. Moore continued to trade until 1899, when the last of the family died. In his book, *Days at the Factories*, James Dodd gives a graphic description of Moore's factory in Clerkenwell. If Thwaites were number one in output, then Moore was number two in the turret clock business. In the 1840s, Moore clocks were still much the same in design as clocks by Thwaites and Reed: they had introduced no innovative features at all. Like Thwaites, they had no excelling skill or knowledge that commended them for making the Great Clock.

John Smith and Sons of Clerkenwell

Like Thwaites and Moore, Smith and Sons carried on the business of making domestic clocks and turret clocks but not in such large numbers as their rivals. We are lucky to have an illustrated description of a visit to their factory that was published in 1851 in *The Illustrated London News*. Smith's turret clock design was very much like those of Thwaites and Moore; it was a standard clockmaker's solution to the need to show the time on an external dial. Smith and Sons did not make large turret clocks and this, together with the fact that they were just basic producers, would have excluded them from any chance to make the Westminster clock.

John Whitehurst III of Derby

There was a succession of clockmakers by the name of Whitehurst working in Derby. The most famous was John Whitehurst FRS (1713–88),

FIG **6.10** A Moore clock of about 1830.

who was also a maker of scientific instruments, an author, and a geologist. John's brother James was also a clockmaker and James had a son, John Whitehurst II (1761–1834). John Whitehurst III (1788–1855) was the son and successor to John II, and he was the last of the clockmaking line.

All the Whitehursts made turret clocks, the majority being supplied in the nineteenth century by John II and John III. In total, the dynasty supplied around 100 turret clocks, so they were well known in this field. They adopted the plan that was used by Hindley of York; this is known as a double-framed clock. An extended base frame contains the long barrels and a shorter frame above that holds the clock trains. This construction gives the

MESSRS. JOHN MOORE AND SONS' CLOCK FACTORY.

DRAWN BY SCHMITTHENNER; ENGRAVED BY H. C. MASON.

FIG **6.11** Moore's factory, as portrayed in an 1852 magazine article. Turret clock making was carried out on the middle and ground floors.

advantage of a long barrel allowing many turns of line and thus giving a longer going period between winding. The shorter frame provided a strong support for the trains of gears without having excessively long arbors, with the attendant problems in assembly.

John Whitehurst III made turret clocks that were of good quality and robust, but did not incorporate any new inventions or unique features. As a well-known maker, it was almost inevitable that Whitehurst should be invited to tender for the Great Clock.

THE TURRET-CLOCK SHOP.

FIG **6.12** John Smith's turret clock workshop.

FIG **6.13** John Whitehurst I, founder of the Whitehurst family of clockmakers.

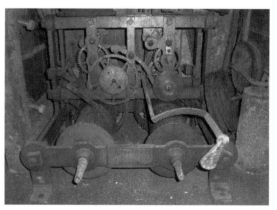

FIG **6.14** A clock by John Whitehurst III.

Edward John Dent

Edward John Dent was born in 1790. He may have developed an interest in watches from his father's cousin, who was a watchmaker. This led to an apprenticeship in watchmaking and then to employment with the famous clock, watch,

FIG **6.15** Edward John Dent (1790–1853). Photo courtesy of the British Horological Institute.

Barracks, Woolwich. A letter written to him from the Admiralty alleging that a turret clock he supplied to Mr G.S. Harrison at Clifton Hampden was *Purporting to have been manufactured by you but which is in reality the commonest description of French manufacture.* Dent responded with a declaration that the clock was made by the workmen in his factory in Somerset Wharf, near the Strand. It was signed by 18 workmen and witnessed by the foreman. Dent had little doubt that Vulliamy was at the bottom of this, as he had also tendered for the job. In a letter to Airy, Dent recounted *I gave two estimates, one for gunmetal wheels for the striking of the hours and quarters, and the other to use cast-iron wheels which would make a saving of ⅓rd the cost in price (I have the patterns by me) and be much better and last an unlimited length of time without repair.*

Overall, E.J. Dent was a self-made clock and watchmaker who not only enjoyed success in his business; he had a scientific bent and was also an inventor and innovator with several patents to his name. In the mid-nineteenth-century, the economy was growing and new churches, town halls, and public buildings all demanded turret clocks. Perhaps Dent saw the opportunity and wanted to expand his business into supplying turret clocks. He may, too, have had an eye on making the Great Clock that was required at Westminster and also the supply of conventional clocks to the new Palace.

George Biddell Airy

George Airy was born in 1801 in Northumberland, of farming stock. He graduated from Cambridge as the Senior Wrangler for his year. Airy stayed at the university, becoming the Lucasian Professor of mathematics and then the Plumian Professor of Astronomy. At the age of just 34 he was appointed Astronomer Royal and started a programme to catalogue stars by their

and chronometer maker, Arnold. Dent eventually went into partnership with Arnold and then founded his own business, becoming highly regarded as a manufacturer of high-class clocks, watches, and chronometers.

Dent was an outsider to the turret clock field but in 1844 supplied a clock for Camberwell parish church and another for the new Royal Exchange building in London. George B. Airy, the Astronomer Royal, was the referee for the Exchange clock and Dent had worked for Airy in supplying clocks to the Royal Observatory. Dent had made a visit to France in 1843 to see turret clocks and for a period up to the early 1850s he sold turret clocks that looked very much like one available from France—but they were copies. In 1845, Dent was tendering for a clock at the Marine

FIG **6.16** An early Dent clock of around 1842 at Caythorpe church. The design is similar to that of French turret clocks.

FIG **6.17** George Biddell Airy, the Astronomer Royal, as a young man.

position. This project involved the installation of the Great Transit Circle that defined the Greenwich Meridian.

Apart from being an expert in astronomy, mechanics, optics, and mathematics, Airy was instrumental in setting up the Shepherd electric master clocks in the observatory, which drove slave dials in all the observing domes. From this came the gate clock and a time signal that was telegraphed to London and hence on to the whole country. Since time was so important to astronomy, he was well versed in clocks. In 1826, he wrote a paper on clock escapement errors and employed E.J. Dent to work on the observatory clocks.

Airy was invited to be a referee for the supply of a turret clock for The Royal Exchange and Lord Seymour invited him to be the referee for the Great Clock at the New Houses of Parliament.

As one of the top scientists of the day, Airy turned his hand to many things from magnetism on ships to the Tay Bridge disaster and making spectacles to correct astigmatism. He was a man who was firm but gentle, and in all things methodical and precise. He died in 1892, at the age of 90.

PLANS FOR THE CLOCK

1844

It was normal for the architect of a large project to be involved in almost all aspects of a building, so it was on 29 March that Barry wrote to

Benjamin Lewis Vulliamy asking him if he would produce plans for the Great Clock and if so, on what terms. Barry had worked with Vulliamy before; St Peter's church in Brighton was one example. Since Vulliamy was clockmaker to the Queen, he seemed the natural choice as maker of the Great Clock. Vulliamy replied on 1 April to say that he could prepare plans for the clock and that it would take him two years to complete. His price would be 100 guineas, but if he was not awarded the contract to make the clock, then the price would be 200 guineas.

Barry duly responded to Vulliamy on 25 April giving him authority to proceed in the preparation of the clock plans. Vulliamy acknowledged the instructions and enquired if the clock should be finished *luxe d'exécution* or just have a functional finish. Barry answered to say that in his opinion all superfluous work in the clock should be avoided.

1845

Vulliamy contacted Barry in January to point out that there had been no mention of producing a quotation for making the clock, but a letter from the office of First Commissioner Earl of Lincoln had alluded to a quotation. Vulliamy pointed out that producing a quotation was not possible until the method for making the clock was settled. Barry had no doubt forwarded this letter to the Commissioners and this must have set them thinking.

Towards the end of the year, E.J. Dent wrote to the Commissioners introducing himself as a maker of clocks and turret clocks, asking to be considered as a candidate for making the Great Clock. The Commissioners replied to say that when the plans and specification were complete, they would include him as a candidate to tender

for the Great Clock and other clocks for the new Palace. Dent wrote straight back to the Commissioners, declining to tender if another person's plans were to be used, saying that it would be detrimental to his reputation and reduce him to a mere executive mechanic. He would, of course, comply with anything under the direction of the Astronomer Royal. Dent appended a letter that he had received from Airy, saying that Airy believed Dent's Royal Exchange clock was the best in the world and that he considered Dent as a most proper person to be entrusted with the construction of another clock of similar pretensions.

HOROLOGY.—WORKING PARTS OF THE CLOCK AT THE ROYAL EXCHANGE, LONDON.

FIG **6.18** Dent's clock for the Royal Exchange, 1844.

1846

In his autobiography, Airy records, 'On 20 June Lord Canning enquired of me about makers for the clock in the Clock Tower of Westminster Palace. I suggested Vulliamy, Dent, Whitehurst; and made other suggestions: I had some correspondence with E. B. Denison, about clocks.' Presumably, letters from Vulliamy and Dent had made Lord Canning and the Commissioners think again about who was going to make the clock.

Neither Thwaites and Reed nor John Moore and Son were considered as possible makers of the Great Clock. The two companies were certainly the largest producers of turret clocks in the country; most probably they were just regarded by Airy as companies that made clocks, certainly their track records showed no invention or innovation of any kind that would be needed to build a Great Clock that would drive four 22-ft dials and strike on a bell of around 14 tons. There were several smaller turret clockmakers in London and the provinces, but no doubt the job had to be awarded to a large company because of the resources and investment needed to complete the job.

Barry responded to Vulliamy's letter of 18 months before: his answer must have horrified Vulliamy, who had probably assumed that the job was his. Barry wrote that the Commissioners had decided to invite tenders for the Great Clock from Vulliamy, Dent, and Whitehurst. Worse still, a specification had been prepared by Sir George Airy that defined some key features of the clock and it recommended that persons wanting to tender would be advised to inspect the turret clock (made by Dent) in the Royal Exchange. Airy's specification was the first of several. The specification follows:

Conditions to be observed in regard to the construction of the clock of the New Palace of Westminster.

I. Relating to the Workmanlike Construction of the Clock.

1. The clock-frame is to be of cast-iron, and of ample strength. Its parts are to be firmly bolted together. Where there are broad bearing surfaces, these surfaces are to be planed.

2. The wheels are to be of hard bell metal, with steel spindles working in bell-metal bearings, and proper holes for oiling the bearings. The teeth of the wheels are to be cut to form on the epicycloidal principle.

3. The wheels are to be so arranged that any one can be taken out without disturbing the others

4. The pendulum pallets are to be jewelled.

II. Relating to the accurate going of the Clock.

5. The escapement is to be deadbeat, or something equally accurate, the recoil escapement being expressly excluded.

6. The pendulum is to be compensated.

7. The train of wheels is to have a remontoire action, so constructed as not to interfere with the deadbeat principle of the escapement.

8. The clock is to have a going fusee.

9. It will be considered an advantage if the external minute hand has a discernible motion at certain definite seconds of time.

10. A spring apparatus is to be attached for accelerating the pendulum at pleasure during a few vibrations.

11. The striking machinery is to be so arranged that the first blow for each hour shall be accurate to a second of time.

III. Relating to the possible Galvanic Connexion with Greenwich.

12. The striking detent is to have such part that, whenever need shall arise, one of the two following plans may be adopted, (as after consultation with Mr Wheatstone or other competent authorities, shall be judged best), either that the warning movement may make contact, and the striking movement break contact, for a battery, or that the striking movement may produce a magneto-electric current.

13. Apparatus shall be provided which will enable the attendant to shift the connexion, by means of the clock action, successively to different wires of different hours, in case it should hereafter be thought desirable to convey the indications of the clock to several different places.

IV. General Reference to The Astronomer Royal.

14. The plans, before commencing the work, and the work when completed, are to be subjected to the approval of the Astronomer Royal.

15. In regard to the Articles 5 to 11, the maker is recommended to study the construction of the Royal Exchange clock.

22 June 1846   *(signed) G.B. Airy.*

Vulliamy's response two weeks later was cool. He declined to tender, objecting to Airy as sole referee, stating that he was strongly prejudiced in favour of an individual known for many years as an eminent maker of marine chronometers, but who had only within the last three years turned his attention to making public clocks. Vulliamy was referring to E.J. Dent.

In August, Vulliamy finally completed the Great Clock plans that he had been contracted to produce; he sent them to Barry, who in turn passed them on to the Commissioners. There were, in all, 36 drawings and the accompanying notes of explanation spanned 20 pages when they were published by order of the House of Commons. The full set of drawings is in the archives of the Institution of Civil Engineers and a partial set is held in the Parliamentary Archives.

Vulliamy's Great Clock was certainly a good-looking design. But Vulliamy so often concentrated on the fine details of his turret clocks whilst really missing the key points of achieving good timekeeping.

In addition to the plans and notes, Vulliamy supplied a further three pages of comments on

Airy's specification. In brief, the points to which he objected were that he...

Objects to wheels made of bell metal, but says he uses gunmetal. (Both of these are generic terms referring to slightly different types of bronze.)

Objects to jewelled pallets, saying he has a preferable manner. (This probably refers to his double-acting self-adjusting pinwheel escapement, where the pallets could swivel in two planes, thus accommodating wear and any irregularity.)

Objects to a compensation pendulum, saying his method of pendulum suspension is better. (He proposed a system of roller bearings that would have had no effect in producing temperature compensation.)

Objects to a remontoire, since marine chronometer makers do not use such a device. (This indicates that he did not really understand the issues with a large turret clock.)

Objects to a going fusee. (Airy did not get his terms right here.)

Objects to the minute hand having a discernable motion at certain definite intervals. (This only happens when a remontoire is employed.)

Objects to a spring apparatus for accelerating the pendulum. (He did not understand what Airy wanted, which was a spring that would speed up the pendulum to bring it to time when the clock was wrong.)

Objects to the first blow of the hour required to be accurate to a second of time. (Dismisses this as impractical due to the speed of sound giving different times depending on where the listener was. In reality, Vulliamy did not have any conception that a turret clock could be made to keep time perhaps 20 times better than the best clocks of that time.)

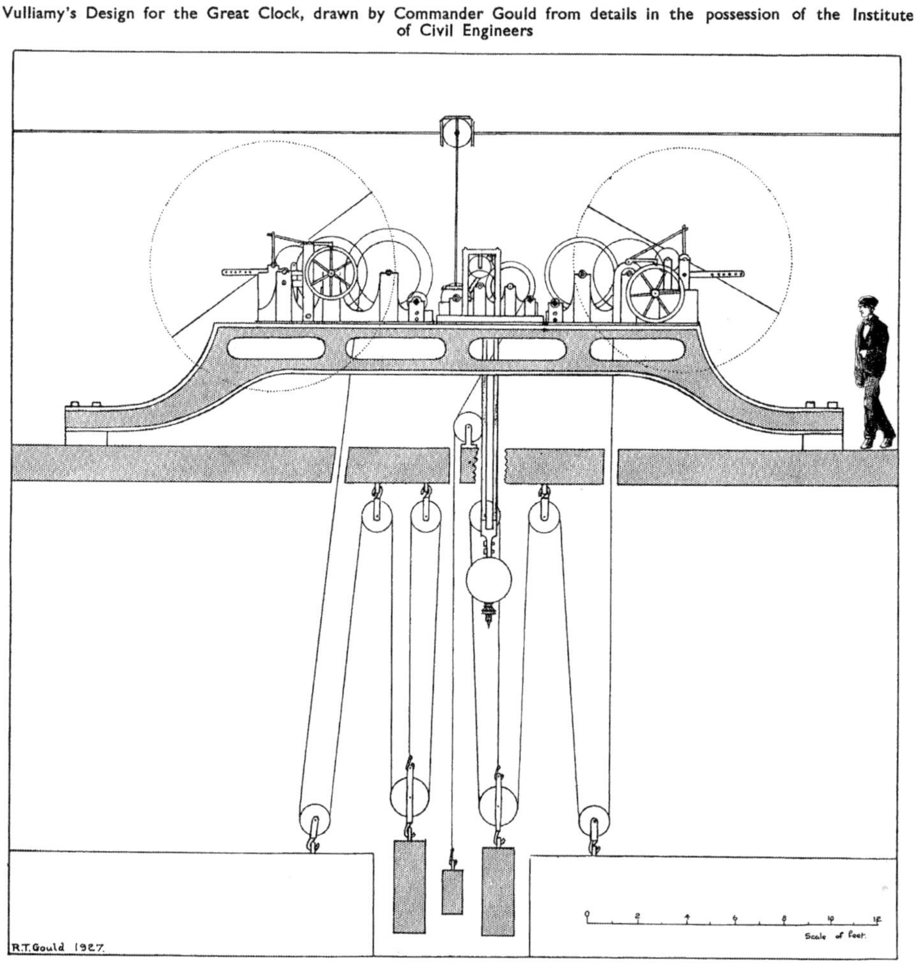

Vulliamy's Design for the Great Clock, drawn by Commander Gould from details in the possession of the Institute
of Civil Engineers

R.T.Gould 1927

Scale of feet.

FIG 6.19 Vulliamy's clock, as sketched by Rupert Gould for the *Horological Journal*. Courtesy of the
British Horological Institute.

Comments on the galvanic apparatus. (Indicates his
unfamiliarity with what is required.)

*Objects to the Astronomer Royal as the sole referee,
and suggests two other names.*

*Declines to see the Royal Exchange Clock until after
he had submitted his plans.*

In all, Vulliamy's objections are a mixture of
not understanding why Airy made certain
conditions and a resentment of Airy's perceived
interference in a field in which Vulliamy thought
he was supreme.

1847

Early in January, the Commissioners replied to
Vulliamy to say that E.J. Dent, in addition to the
plans he submitted, intended to supply a quarter
size cardboard model of the clock, and they
enquired if Vulliamy would also submit a model
of his proposed clock. Vulliamy wrote quickly

back to them saying that he would consult with some friends; a week later he told the Commissioners a cardboard model was, in his opinion, of no use. He was, however, making a turret clock for Mr Peto and this was a little larger than quarter size and '*I am purposely making it as like the clock I planned for the Houses of Parliament as it is practical to make a small clock like a large one.*' The Commissioners were invited to inspect this clock as soon as it was finished.

Mr Peto was the builder of the Houses of Parliament; he had bought for himself Somerleyton Hall in Suffolk, which he was in the process of extensively remodelling, including the addition of a clock tower. Somerleyton Hall's clock tower contains a three-train clock by Vulliamy made on his flatbed principle. Comparing this movement to what Vulliamy proposed for Westminster reveals that it is just one of Vulliamy's standard flatbed movements with none of the features he intended to use at Westminster. A brass plate fixed to the clock records that the clock was a scale model of what Vulliamy planned for Westminster. Peto was later knighted, becoming Sir Samuel Morton Peto.

Airy reissued his specification in February, correcting condition number 8 from going fusee to going barrel. He emphasized that condition number 11—that the first blow of the hour bell had to be accurate to a second of time—would not be relaxed. Altered or added items were reported as follows:

12 The striking detent or striking hammer for the hours is to be so arranged that at every blow it will break contact with a powerful magnet, mounted on the principle recommended by Mr Wheatstone, for the formation of a magneto-electric current.

And the following condition is new:

16 The hour wheel is to carry a ratchet shape wheel or a succession of cams, which will break contact with a powerful magnet, on the principle recommended by Mr Wheatstone, at least as often as once in a minute, for the purpose of producing a magneto-electric current, which will regulate other clocks in the New Palace.

As the new condition No 16, requiring increased power in the maintaining force, will probably make it necessary that the barrel of the first wheel of the going part be somewhat increased, so as to take full advantage of the depth allowed for the fall of weights, it is requested that (if it is judged necessary) sketches may be sent to the Office of Woods and Works, showing generally the alteration which this will make in the plans already sent.

And as the new condition will slightly increase the expense, it is requested that final tenders, applying to the last plans sent to the Office of Woods and Works or to the Astronomer Royal, and embracing every part of the machinery (stating distinctly whether the iron or stone supports, the clock·hands, and the clock dials are included), be sent to the Office of Woods and Works on or before Monday, 16 March.

(signed) G.B. Airy.
11 February 1847.

Vulliamy responded with a Statement on the Grounds on which B.L. Vulliamy claims to be employed to make the Great Clock for the new Palace of Westminster. He went through his experience with turret clocks, connections with the Crown, and included testimonials. An explanation was given on his new mode of construction

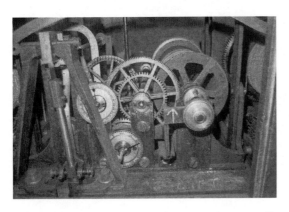

FIG **6.20** Going train of the clock at Somerleyton Hall. Reproduced by permission of Lord Somerleyton.

of a turret clock, giving sizes and weights, but never any mention of timekeeping performance. Famous clocks he had made were listed and in conclusion he stated that he used the factory of the Bramah brothers to make his turret clocks.

Estimates were received from both Dent and Whitehurst. Vulliamy did not tender. The estimates and plans were passed to Airy, who considered them and reported back to the Commissioners in May. Airy analysed the two submissions carefully dividing his report into categories. The following is a summary of his report.

1 On the means possessed by Mr Dent and Mr Whitehurst for making a clock of very large dimensions

Both makers were reported as having the means to make large clocks, with Dent's tools being considered somewhat superior to Whitehurst's, since he had larger lathes and a planing engine.

Both makers would need considerable assistance from an external engineer for the large frames and the great wheels.

2 On the personal abilities of the two makers

Dent is noted as a clockmaker who was used to paying particular detail to accuracy and is aware of timing and rating, since he was a chronometer maker. He had travelled to France to see turret clocks and had recently engaged in turret clock-making. Whitehurst was a traditional and enthusiastic clockmaker familiar with English clocks and turret clocks only in his locality.

If it were necessary to entrust the making of the clock, without any control, to one of other of the two makers I should prefer Mr Dent; because I think it is easier for Mr Dent to acquire Mr Whitehurst's solidarity, than for Mr Whitehurst to acquire Mr Dent's accuracy. But, under the most trifling control, either of these makers will certainly construct the clock in a perfectly satisfactory way.

3 On the plans that have been sent in

I had difficulty in obtaining explicit plans form Mr Whitehurst. Mr Dent's plans, it is understood, are not final.

Airy goes on to explain that the provision of detailed plans being a great expense and labour, the submitted plans were sufficient.

Airy now had doubts about electrically driving other clocks from the Great Clock, following experiments he had conducted, but he did add the comment, '*The transmission of a signal to Greenwich by a magneto-galvanic current is a matter of no difficulty, and I recommend that preparation be made for it.*'

4 On the sums demanded by the makers for the construction of the clock

Mr Dent   £1,600
Mr Whitehurst   £3,373

In explaining the great price difference, Airy considered that

Whitehurst had made his estimate at what might be called a 'paying price' while Mr Dent, who is one of the most enterprising of London tradesmen, has not improbably been willing to construct the clock at a loss to himself for the sake of reputation which he hopes thereby to acquire.

Airy concluded by saying that he was not competent to make the final decision and that he must leave this to the Commissioners.

In a separate report of the same date, Airy reported on Vulliamy's plans; in the main, he dealt with Vulliamy's complaints of Airy as sole referee, pointing out the Great Clock was just a large clock, demanding no new principles, and furthermore Vulliamy was often asked to give advice to makers of clocks, particularly chronometers.

When Airy came to the plans he wrote

I have very carefully examined Mr Vulliamy's beautiful plans. In regard to the provisions for strength, solidarity,

power, and general largeness of dimensions, they are excel-
lent. In regard to delicacy they fail; and they fail so much,
that I think myself justified in saying that such a clock would
be a village clock of very superior character, but would not
have the accuracy of an astronomical clock. I do not assent to
the advantage of Mr Vulliamy's peculiar pendulum suspen-
sion. I know that it would fail for a balance, or for a vertical-
force magnetometer, and I believe that it would fail for a
pendulum. I do not assent to the advantage of Mr Vulliamy's
turning pieces for pallets and hammer tails, believing that
they will introduce evils more important than those which
they are intended to remove.

And so Airy damned Vulliamy's plans, which
had taken two years of work and was to cost the
Commissioners the 200 guineas originally
quoted.

THE TENDERS

In 1846, John Whitehurst III submitted his plans
to the Commissioners; they were dated 24
September. In a covering letter, he mentioned
the clock at Bray that he had just installed in the
parish church. Visitors to the clock today can see
a double-framed movement with a pinwheel-
type escapement. The clock is a well-built
two-train movement. A nice touch is a small

brass spanner, which serves to adjust the octagonal rating nut that is on the top of the pendulum
suspension. There is also a double-ended spanner
that was provided with the clock to work on the
movement.

Whitehurst's drawings covered three sheets;
these are in the National Archives and show a
large double-framed movement. Great wheels
were about three feet in diameter and the frame
was about nine feet long. The escapement was a
pinwheel: there was no remontoire to address
the problem of varying power caused by wind
on the hands. Various questions were put to
Whitehurst and, early in February 1847, Airy
visited Derby and saw Whitehurst at his factory.
Airy recommended that Whitehurst visit the
Royal Exchange clock in London. One wonders
what Whitehurst thought of this; Dent was the
maker and Airy had a hand in its design.

1847

On 3 May 1847, Dent wrote to the Commissioners asking that he be allowed to tender for
other clocks in the Palace. Having had no reply,
he wrote again in June and again he received no
reply. This was obviously a good business

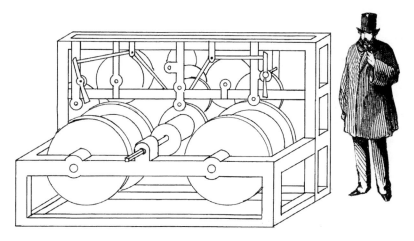

FIG **6.21** Whitehurst's design
for the Great Clock, derived
from plans in the National
Archives.

FIG **6.22** Whitehurst's clock at Bray church, 1840.

opportunity; with something like 1,000 common dial clocks needed, that represented a significant income on the back of his loss-making great Clock. On 3 July, he wrote yet again; this time complaining about how Vulliamy was supplying clocks for the House of Lords. In protest, he stated he would withdraw from making the Great Clock and demanded that his plans be returned. The implications for Dent were simple, it was probable that if he made the Great Clock it would be at no profit or at a loss, but there was a good potential for supplying something like 1,000 dial clocks to the Palace. A dial clock sold in the trade for around £5 to £10, depending on size and finish. Overall Dent would be financially secure if he could secure such a contract.

The Commissioners responded in July to say that no clocks had been ordered for the Palace but that Barry had ordered a consignment of furniture for the new House of Lords, which included clocks made by Vulliamy; the bill had been sanctioned by the Treasury. Barry had now been ordered not to purchase any further clocks. Regarding the request for return of Dent's plans, this was refused for the time being, as they were now registered official documents.

Thwaites and Reed, who had remained silent throughout all the proceedings, got wind that Dent wanted to withdraw and they wrote to the Commissioners in late August requesting that they be allowed to tender for the Great Clock. In support of their claim, they said that the company had been established for over 100 years and made over 3,000 turret clocks.

The Commissioners wrote to Dent to say they were willing for him to remain a competitor, and also consulted with Airy as to his opinion of Thwaites and Reed. Airy did not see any advantage in admitting Thwaites and Reed, so the Commissioners replied to Thwaites saying that they were under a misapprehension and that nothing had changed regarding the competition for the Great Clock.

In his autobiography, Airy reported that in May 1847 he had examined and reported on the factories of Dent and Whitehurst. Here he states that '*Vulliamy was excessively angry with me.*'

In July, Edmund Beckett Denison Senior, the MP, moved for all the papers regarding the Great Clock to be tabled, then the House of Commons ordered the Return to be printed and the documents passed into the public domain. These papers covered the period from 1844 up to July 1847. No one cried *Nepotism*, since Edmund Beckett Denison Junior, despite his interest in public clocks, had not yet appeared in the public eye. It was very probable that Edmund Junior had discerned that the Great Clock was potentially heading for disaster. Wanting to know the full story of what was going on behind the scenes, he could well have asked his father to have the papers tabled. Perhaps in his self-confident manner, he may have thought that he would be able to manage the whole situation. Oddly enough, some years later, when Edmund Junior ended up as the sole referee and designer of the Great Clock, no one cried *Nepotism* either. In 1848, the MP Mr Wyld asked for more papers and again the House of Commons ordered for them to be printed. For completeness, the House of Commons in June 1852 ordered that the two sets be reprinted.

Edmund Beckett Denison

Edmund Beckett Denison was born in 1816. His father, also Edmund Beckett Denison, was a businessman, who worked in Doncaster. Edmund Junior went to Doncaster Grammar School and then to Eton College. Before going up to Trinity College, Cambridge, he was tutored by the Rev. Hodgson at Helions Bumpstead in Suffolk.

These were two highly formative years of his life; years later (about 1856) Edmund wrote:

The first year we were there, we started bell ringing at my suggestion & got a subscription up for recasting two of the bells which was cracked, and the next year we got a 6th bell added. We used to go about the country to see and ring the bells. The two recast and the new one were hung by me and the village carpenter. I used to spend near all my time there out of reading hours in carpentering & after mechanics. I had little expectation then that my clockmaking & bell ringing would bear much fruit 20 years afterwards.

Edmund read the mathematical tripos at Cambridge, graduating in 1838 as 30th wrangler, not very far from the bottom of the class. Whilst at Cambridge, he frequently attended St Mary's church, where he was a bell ringer. The chime of St Mary was as familiar to Edmund as it was to George Airy. On leaving, he recorded '*In August 1838 I put up the church clock in Bumpstead, the first I had to do with. It was made by Vulliamy of London, then the best maker.*' It was also recorded that he put up the clock with the aid of the village carpenter. In 1839, the church paid Abraham Wright £12 7s. 8d. for carpentry work about the church. This must have been the man with

FIG 6.23 Marble bust of Edmund Beckett Denison in his later years.

FIG 6.24 Clock by Vulliamy installed in Helions Bumpstead church by Edmund Beckett Denison and Abraham Wright.

whom Edmund spent so much of his time at Helions learning mechanics. It is to this obscure village carpenter that we owe a great debt for his input in the Great Clock of Westminster.

After graduating, Edmund read law for a year at Cambridge and then went to London where he studied to be a barrister. In 1841, he passed his MA and was called to the Bar. At first, he took on Chancery work but being only a junior in a complex case he was forced to look for other work and, on the suggestion of a colleague, tried Parliamentary work. Here he was the ideal person in the right place at exactly the right time. The railway boom had just started and, shortly afterwards, his father had become the MP for Doncaster. As MP, Edmund Senior was invited to be the Vice Chairman of the Great Northern Railway Company; and since the Chairman was ineffective, he was essentially the Chairman. Edmund Junior soon established a fearsome reputation as a lawyer framing private bills to get railways lines set up across the country. He made a lot of money and eventually became a millionaire.

As an amateur horologist, Edmund went on to design the Great Clock single-handedly and the bell Big Ben. He was never far from controversy and had few friends.

1848: Lord Brougham and the House of Lords' Papers

Henry Brougham was born in Edinburgh in 1778, attended Edinburgh University at the age of only 14, and later read law. He was a keen journalist and had a scientific nature, becoming elected a Fellow of the Royal Society. Henry chose to follow law and was admitted a member of Lincoln's Inn and called to the Bar. He had a keen social conscience and entered into Whig politics; he was actively involved with the abolition of slavery and the Reform Act of 1832. In 1830, Henry was elevated to the Peerage as Lord Brougham and Vaux.

Brougham was always a Radical and a champion of good causes. He was, in the main, a popular man, regarded as fair and honest. During 1847, the House of Lords had various papers tabled concerning the Great Clock and in February 1848 they were ordered to be published. In the main, the letters were the same as those that were printed by the House of Commons. Lord Brougham was the leading force in publicizing these papers: quite probably Vulliamy had approached him with his grievance about the way things were going concerning the clock. Vulliamy's political view was probably far from that of Brougham, but Brougham would have been pleased to act if there was a possibility of an injustice taking place.

By early August, Dent had changed his mind about withdrawing, probably the result of having seen the official release of papers relating to the Great Clock as moved by Lord Brougham in the House of Lords. From this, he concluded that information had been leaked to Vulliamy from the Commissioners Office.

Some of the House of Lords' papers were reprinted under the wordy title A Portion of the Papers Relating to the Great Clock for the New Palace at Westminster, With Remarks, for Private Circulation Only. It was printed by William Clowes and Sons, Stamford Street, in 1848 and there can be little doubt that Vulliamy was the author. Various annotated copies survive that bear Vulliamy's signature and the name of the person to whom the publication was sent.

The booklet's introduction concludes:

A careful perusal of the papers, in the order in which they are here presented, irresistibly leads to the following conclusions:

1st. That Mr Airy was, throughout, not only very strongly biased in favour of Mr Dent, but had, from the first, made up his mind that Mr Dent, and no one else, should make the clock.

2nd. That Mr Vulliamy was, under the circumstances here detailed, fully justified in declining to compete, if Mr Airy were to be constituted sole referee, and all proceedings connected with the clock placed under his absolute control.

If these conclusions appear to be justified by the contents of the correspondence, perhaps it may not be too much for Mr Vulliamy and Mr Whitehurst to desire that the whole of the papers furnished by the three competitors be referred to a committee of scientific men, for their opinion and report.

More Clockmakers

Early in 1848, two further applications were received by the Commissioners from turret clockmakers requesting to be allowed to tender for the Great Clock. One was from an unknown clockmaker, Samuel McClellan of Greenwich, the other was James Mangan of Cork in Ireland. Both requests were refused.

Denison wrote to First Commissioner Lord Morpeth in May, reminding his Lordship that he and his father visited him last year. He then presented himself as knowledgeable on public clocks and stated that Vulliamy was in concert with Barry to get Dent ousted from the job of making the Great Clock. He requested a meeting but the outcome is not known. It does however indicate that Denison was squaring himself up for involvement with the Great Clock.

In August, Vulliamy requested that he be paid the 200 guineas for the plans he had prepared for the Great Clock. He was paid 100 guineas only. The Commissioners said that since a maker for the clock had not been chosen, they could not advance the other 100 guineas that were contingent on him not being chosen to make the clock.

London news was often included in provincial newspapers. In January, the *Chester Chronicle* printed a column on the proposed Westminster clock, giving full details based on what had been published in Parliamentary papers. An interesting note was that wire rope was to be specified; the columnist stated that this was used on the Royal Exchange clock and that the manufacturer stated that a ½-inch diameter wire line could hold 18 cwt without breaking. Cable-laid catgut had also been discussed; no doubt this was a proposal from Vulliamy, who used catgut lines on a clock installed in 1829 in Windsor Castle. In his book *A Rudimentary Treatise on Clocks, Watches, and Bells* published in 1850, Denison mischievously wrote

In the Windsor Castle clock Mr Vulliamy used catgut ropes (wire ropes were not then invented); but they are enormously expensive: he was told that 17,000 sheep contributed their entrails to compose these catgut lines.

1849

Vulliamy felt compelled to request his outstanding 100 guineas in May, presenting as his case that he was now 70 years old and that he doubted if his life would be prolonged sufficiently as to make the clock. Also, work on the Clock Tower had advanced slowly and progress had for some time been almost stationary. Vulliamy did not relinquish his claim to make the clock but suggested that the 100 guineas be taken on account, if it be determined that he should make the clock. The Commissioners replied that there was no immediate prospect of choosing a maker for the clock, so his request for payment was refused. Not to give up, Vulliamy appealed to the Special Commissioners for completing the New Houses of Parliament. A dry response said that it did not appear to the Commissioners that there was any probability at the present of any decision being required relative to the person to be employed to make the clock. Vulliamy was financially secure and the second payment, though nice, was not essential to him. Quite possibly he would have been granted the payment if he gave up his claim to make the clock on the grounds of advancing years and poor health. However, it looks like his pride did not allow him to let go of the prospect of making his masterpiece.

One William Waldron had made a proposal to the Commissioners concerning the striking of the Great Clock. No details remain but it seems probable that he had made the suggestion of a warning blow to be sounded before the hour to alert people to listen, or repeating the hour a little after it had been struck, in the manner of continental clocks. A curt order was issued to return the plans to its author.

1850

The year opened with the publication in January of the book *A Rudimentary Treatise on Clocks, Watches, and Bells* by Edmund Beckett Denison. As such, it was the first book in English with extensive information on turret clocks. Two of Denison's aunts lived in Meanwood near Leeds. One of their gifts to the church was a clock and their clock-loving nephew was engaged to provide the design. The frontispiece of the book shows the Meanwood church clock movement, which was made by E.J. Dent.

There are several features interesting about this clock. It follows the general flatbed design, has a spring remontoire to keep a constant force to the escapement, and has a compensation pendulum. All the clock dials are concave instead of convex, a feature that Denison claimed reduced the viewing error due to parallax. Denison also favoured batons on the dials instead of Roman chapters, claiming that everyone knew where the numbers were.

A Rudimentary Treatise on Clocks, Watches, and Bells went to eight editions, the last being published in 1903. From the second edition onwards the frontispiece was the Westminster clock and there was a chapter on the history of the Great Clock, told, of course, from Denison's viewpoint.

1851: Winding of Clocks

It was a normal practice for any large organization, country house, or mansion to have its clocks wound by a clockmaker. Generally the clockmaker or his assistant would visit the building once a week, wind all the domestic clocks, set them to time, and make any necessary minor adjustments. Any clock found to be faulty would be taken away, repaired, and returned a few weeks later. Where there was a turret clock, this too was wound; if the clock was old, it was possible that it had to be wound every day. Clock winding was a service that was charged for and contracts were entered into. Today clockwinding is still carried on, mostly in London Government offices and the Royal Palaces.

The Office of Woods and Forests, etc., was responsible for the winding, care, and supply of clocks in Government buildings. Vulliamy had a contract for winding Government clocks and this included many turret clocks, such as for the Horse Guards and the Post Office.

In a flurry of activity that only spanned March and April, The Office wrote to Messrs Vulliamy, Dent, Blundell, Dutton, Frodsham, Gaze, MacDowall, Joyce, and Whitehurst, inviting tenders for the care and supply of clocks. A specification was provided to cover the supply of new clocks, as well as the winding, regulation, cleaning, and repair of existing clocks. The contract would run for five years. For new clocks, various types of spring and weight clocks were listed: details were given of dial sizes and the wood of the case was defined. The whole was couched in terms that were faintly legalistic; one might suspect that Denison had an input but another person had then added other requirements. Very much in Denison's style was '*Escapements to be recoiling unless some new escapement equally good can be applied without increasing the price.*' Another similar requirement was '*The reverse minute wheel should be set on a cock with pivots and not on a stud.*' However, the sizes of driving weights were specified, as well as the weights of the pendulums along with the requirement that pendulums should be brass cased...all things Denison would certainly not have had a hand in.

Vulliamy responded in indignation, saying that he had cared for Government clocks for over 50 years, he supplied various references and stated that he often loaned clocks at no extra charge. It appears that he did not tender; his attitude was that he had done the job for 50 years and resented the new ways. He objected to the clause that the contract should be for five years, quoting his advanced years and poor health.

The Office was not interested; they requested that Vulliamy hand over all the keys to the clocks and he was thus ousted from his long service. Airy once wrote of Vulliamy '*A difficult man of unmanageable temper.*' No doubt the Office was pleased to be in control of their business and not have the ageing Vulliamy to contend with. Mr Dutton of Fleet Street was awarded the contract for the care of all the public clocks. It is not known who won the contract for the supply of clocks, mostly domestic dial clocks for use in offices; neither is it clear if such clocks were to be supplied to the New Palace of Westminster.

Vulliamy's final act was to publish a pamphlet *A Statement on the Circumstances Connected with the Removal of B.L. Vulliamy* by the Commissioner of Woods from the Care of such Government Clocks as are in their Custody. In its 24 pages, we see a hurt and ageing Vulliamy clutching onto his traditional approach, unable to accommodate new ways.

Dent's Clock at the Great Exhibition

The Great Exhibition in Hyde Park was the major event of 1851 for the country. It opened on 1 May and around six million visitors had passed through the doors by the time it closed on 15 October. Every aspect of art, agriculture, and industry was displayed, including turret clocks. Dent exhibited a three-train flatbed movement; it was designed by Edmund Beckett Denison and had a remontoire. Once the Exhibition finished, the clock was installed in King's Cross Station, the London terminus of the Great Northern Railway, of which E.B. Denison's father was chairman. At some time in the twentieth century, the clock movement was replaced by an electrical movement and the mechanical movement dismantled.

Little happened as regards the Great Clock until the end of August, when Airy wrote to the Commissioners requesting two sections of the clock tower and a plan view of where the clock was to be mounted. Airy also asked if the weights might be dropped down the air shaft and that he could be considered in communication with the architect. His request to communicate directly with Barry was not granted, but Barry was contacted via the Commissioners. Barry then produced a section and a plan saying the second section was unnecessary; these were then passed to Airy via the Commissioners. Barry also disclaimed knowledge of the air shaft and referred his enquirer to Dr Reid, the heating and ventilation specialist, so the Commissioners had to engage him in correspondence.

This series of communications is particularly interesting, since they show first the bureaucracy imposed by the establishment in maintaining a hierarchical communication network. Second, they indicate that either Airy did not know much about the tower size and construction, or that Barry was lax in issuing information. It must be recognized as a major shortcoming that some key information relating to the clock was not known at such a late state of the proceedings.

Whitehurst called on Airy in October to discuss some aspects of the Great Clock. From this it emerged that no one had thought of the need for having someone to look after the clock after it had been built. Airy mentioned that he was in communication with the South Eastern Railway and the Electric Telegraph Company concerning a telegraph line for transmitting a signal from Greenwich to Westminster at 30 minutes past each hour, and a signal from Westminster to Greenwich for the purpose of monitoring the clock's performance. He also asked for the authority for himself, Denison, and any other person to proceed regarding the Great Clock.

Airy reported to the Commissioners on 18 May, recommending that Dent be appointed to make the Great Clock. Whitehurst quoted £3,373 and Dent £1,600. He stated that, 'It would be easier for Dent to acquire the solidarity of Whitehurst than Whitehurst to acquire Dent's accuracy.'

A most significant event happened at the end of 1851. Airy's autobiography records:

In November I first proposed that Mr E.B. Denison should be associated with me. About the end of the year, the plan of the tower was supplied to me, with reference to the suspension of the weights and other particulars.

From this point, Denison had a significant input on the design of the Great Clock. Edmund moved quickly and the plans for the clock that were published at the end of January must have incorporated many of his ideas.

The year's correspondence closed with a communication from Airy to Lord Seymour seeking permission to establish the telegraph line connection to Westminster.

1852

Airy and Denison wrote to Lord Seymour following a meeting:

> *42 Queen Anne Street,*
> *29 January 1852*
>
> My Lord,
>
> *We now send, as you requested, two drawings and a short specification for the great clock, which we believe will be sufficiently intelligible to Mr Dent. The specification is prepared also in some respects with the view of explaining to your Lordship, without further trouble, a few points, in which we think provision ought to be made in the tower for the person who is to make the clock.*
>
> *It may be convenient to your Lordship that we should repeat what we said in our interview with you last Tuesday; that if Mr Dent should agree to make the clock according to this new plan, for a sum not much exceeding his former tender, we believe it will be most advantageous to the public service to accept his offer. For even if the question should be re-opened, and a new competition invited, and if any maker should offer to make the clock for a lower price than Mr Dent's published tender, for the sake of the reputation he would thereby acquire, we do not think there is any other person who is actually able to make a clock so different in construction from those of all the other makers except Mr Dent; at least, it would require such an amount of instruction and constant personal superintendence from us as we could not give, not to mention the obvious unfairness which there would be in our teaching other persons how to copy the contrivances and construction which Mr Dent has introduced or hitherto alone used, and some of which must be adopted in this clock if it is to have the character which your Lordship and the public will expect.*
> *We remain &c.*
> *Signed*
> *G B. Airy*
> *E B. Denison*

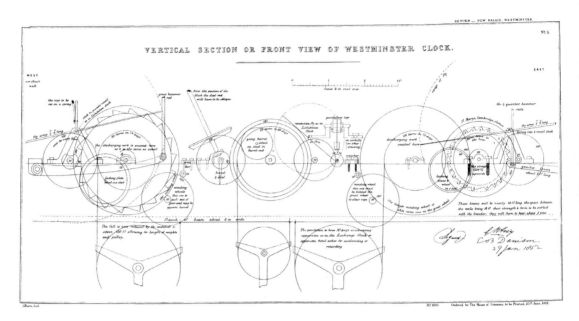

FIG 6.25 Front view of the Great Clock.

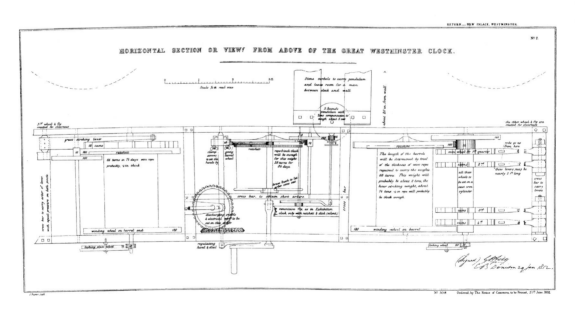

FIG 6.26 Plan view of the Great Clock.

The plans signed by Airy and Denison show a clock that is not the same as the clock that is in the tower today. Differences are that: the flies for the hour and quarter striking are situated behind the clock; the going train has a remontoire; and the design layout of the striking trains is different. However, the general size and construction is similar to the final clock.

The probability is that Denison and Airy had a meeting in Denison's house at Queen Anne Street. The above letter to the Commissioner was composed and a set of plans signed by Denison and Airy was attached. A set of drawings in the Royal Greenwich Observatory archives has plans of the clock in sketch format; they are dated 29 January 1852 and signed E.B. Denison and the comments are in his handwriting. They are not signed by Airy.

In the manuscript collection of the Worshipful Company of Clockmakers, there are two plans that came from Dent. These are the same as the plans published in the Parliamentary Return, are signed by Airy and Denison, and are annotated in Denison's hand.

Plans and the specification were sent to Dent who replied that his estimate would be £1,800 and the clock would be completed in two years. On 25 February 1852, the Commissioners wrote to Mr Dent, accepting his offer and giving him authority to proceed with the making of the Great Clock. Barry was instructed to provide any information requested to Dent and Airy. All the indications are that Edmund Beckett Denison had designed the clock himself.

DENT ACCEPTS

Dent acknowledged the command to make the Great Clock on 8 March. He requested that he retain the two drawings for the present along with others containing corrections. From the time that Barry first made an enquiry of Vulliamy, eight years had elapsed before Dent had been contracted to make the clock. As to what Dent's original plans for the clock were is unknown, since his plans have been lost.

CHAPTER SEVEN

THE GREAT CLOCK IS BUILT: 1852 TO 1854

1852

Dent Accepts

After receiving the official order of 8 March 1852, Dent acknowledged the order and commenced work. He had just two years to complete the Great Clock.

The Clockmakers' Company *Memorial*

Vulliamy was several times the Master of the Worshipful Company of Clockmakers and, as such, he was a respected senior liveryman. No doubt Vulliamy was incensed when Dent took the prestigious contract for the Royal Exchange clock and much more so with Dent's having won the contract for the Great Clock. The Clockmakers' Company rallied behind Vulliamy and, in a last-ditch attempt to confuse the contract, produced a verbose petition, a *Memorial* to the Commissioners. Their argument was that the latest specification was different from the original one devised. They also commented that Dent was less experienced than Vulliamy or Whitehurst, cast doubt on Denison, since he was a QC with no practical clock-making experience, and raised the old issue of cast-iron wheels, stating that these were much inferior to those made of gunmetal. They concluded by requesting that if the order had definitely been placed, then a committee of referees be formed to monitor progress and advise on the Great Clock. The committee should contain Barry and several engineers, such as Sir John Rennie.

Although he objected to cast-iron wheels, Vulliamy did indeed use them in some of his turret clocks. As early as 1847 he was using cast iron for great wheels and running steel barrel arbors directly in cast-iron frames. Perhaps he was trying out Dent's scheme to see if it worked.

FIG 7.1 Bearings of the Clockmakers' Company.

During his lifetime, Vulliamy supplied around 100 turret clocks, more then Dent did in their working period. Vulliamy's clocks were very well made and have stood the test of time.

The whole *Memorial* was a severe case of sour grapes: Dent was not in the Company, Vulliamy had lost the job, and the proposed referees had worked with Vulliamy in the past. The Clockmakers' Company was locked in tradition and had failed to realize that together Denison and Dent were a powerful pair of innovators.

In May 1852, a response from Lord John Manners, the Commissioner, was sent to the Worshipful Company, stating that although the order had been placed before he came into office, he considered it his duty to ensure that it was fulfilled; neither would he enter into any controversy nor express an opinion. He did, however, say that he would consider a committee of referees. The committee of referees was never set up.

E.J. Dent and Turret Clocks

Edward John Dent had a clock on the front of his shop made by Paine of London, and it was reputed that the poor performance of this clock drove him into taking an interest in turret clocks. When Dent was awarded the contract for the Great Clock in February 1852 he was an established turret clockmaker. However, he was not a prolific producer. A small factory had been set up in Somerset Wharf near The Strand, in what Denison described as a *stable yard*.

William Badderly was involved with bells. In commenting in *The Mechanics' Magazine* on the exchange of letters between Loseby and Denison in 1857, he made some observations on Dent. He wrote:

I remember that when a clock was wanted for the new Royal Exchange, it was ordered from a watchmaker who had neither the workshop, tools, workmen, nor experience necessary for the occasion. However, the workshops were built, the tools were provided, and the workmen obtained, by tempting them with constant employment at higher wages. The skill of the workmen, however, did not extend much higher than their fingers, and the experience had to be bought; and, according to a familiar proverb, the dearer the better!

The result, in this instance, of the 'prentis hand', like that of many other amateur performances, fell much below the mark; and the unlucky bellfounder was, to some extent, made the scapegoat. At any rate, the Royal Exchange clock has been the subject of so much bitterness that it has now become by common consent a tabooed topic.

'Oh, no we never mention It, the chimes are never heard.'
'Our lips are now forbid to speak that one familiar word.'

Just how truthful these comments are we cannot tell; however, at the least it looks as if Dent created a workshop in 1844 specifically for making the Royal Exchange clock. A report said that it was well equipped with large lathes and a planing machine. As to turret clocks made previously by Dent, it was certainly possible to have any clock, complete or in part, made in Clerkenwell, and the real maker had no issue in putting another's name on the finished item.

Before 1859, there were four very different types of turret clock signed by Dent. First there are the French-type clocks. It was in July 1843 that Dent journeyed to France to see turret clocks; he might have met up with A.P. Borrel, (1818–87) who was a pupil of Wagner of Paris and a maker of turret clocks. There is little doubt that before 1843 Dent either imported Borrel clocks or, most probably, bought one and copied it. In 1841, Dent supplied a two-train clock of the Borrel pattern to Caythorpe church in Lincolnshire, the gift of Colonel George Hussey-Pack. Other clocks of the Borrel pattern followed at Woodbury (1846), Tavistock (1849), a private house in Norfolk (1851), and at Lode and Northchurch (1853).

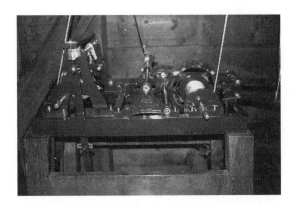

FIG 7.2 French turret clock by Borrell.

These clocks were all two-train movements and had pinwheel escapements. In France, the pinwheel used was of quite small diameter, Dent used a relatively large pinwheel. All the pendulums were compensation using Dent's early style, but the Woodbury clock of 1846 employed the Great Clock's type of compensation with concentric tubes. These characteristics, plus the fact that all movements are signed Dent in the A-frame castings, suggest that they were indeed made by Dent. After 1853, Dent seems to have stopped using the French-style movement.

In 1844, Dent made two turret clocks of the double-frame variety; this is his second type of turret clock. One clock was made for the Royal Exchange and the other for the church of St Giles in Camberwell. The Royal Exchange installation, which included a carillon, was not a success. There were endless problems with the bells and the clock escapement, with its remontoire, did not perform well. The Royal Exchange clock was replaced early in the twentieth century, but the St Giles clock remains.

Two special movements were designed by Edmund Beckett Denison; these can be called Dent's third type. The first was made in 1850 for Meanwood church near Leeds; it was donated

by two of Denison's aunts who lived in Meanwood Hall. This clock is of the flatbed design; it has a pinwheel escapement and a compensation pendulum. A spring remontoire was incorporated that was released every half minute. Denison was, no doubt, trying out his ideas of improving timekeeping. Today, the clock is still in the church but the remontoire has been removed. Records of Potts of Leeds, who maintained the clock, reveal that there had been many problems with this clock and the solution was finally the removal of the remontoire. In the first edition of his book *A Rudimentary Treatise on Clocks Watches and Bells*, Denison uses an engraving of his clock at Meanwood for the frontispiece.

In 1851, Denison designed a three-train clock for Dent, for the Great Exhibition of 1851. In the official catalogue for the Great exhibition, Dent is recorded as exhibiting a variety of clocks, chronometers, and watches with no specific reference to any item. In the catalogue, there is a record of '*A turret clock by E.J. Dent striking on a bell of Union Metal*' in the Main Avenue West. Like the Meanwood clock, this movement had a compensation pendulum and a remontoire; the fly can clearly be seen in an old photograph. Dent was awarded the Council Medal by the Exhibition Committee for his large turret clock on occasion of '*Its strength and accuracy of timekeeping attained in it, which are also accompanied by a cheaper mode of construction than other turret clocks of high character.*' Denison came in for criticism for the award, since he was the Chairman of the Jurors that gave the medal. His response was that he did not propose the medal and said nothing during the Jury's discussions; remaining silent was something he was not used to.

In later correspondence, an attempt was made to ridicule Denison's claim for the excellent

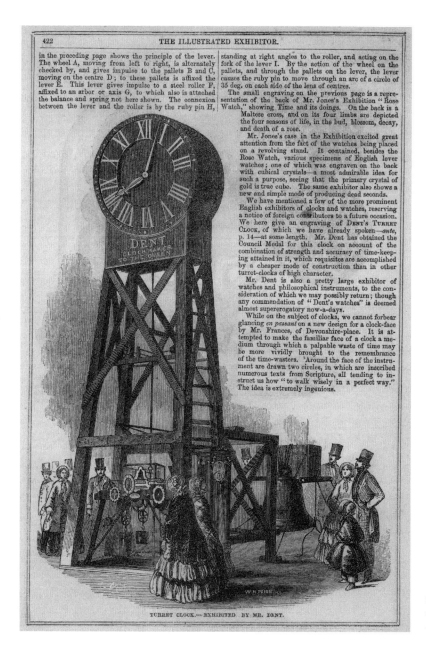

422   THE ILLUSTRATED EXHIBITOR.

in the preceding page shows the principle of the lever. The wheel A, moving from left to right, is alternately checked by, and gives impulse to the pallets B and C, moving on the centre D; to these pallets is affixed the lever E. This lever gives impulse to a steel roller F, affixed to an arbor or axis G, to which also is attached the balance and spring not here shown. The connexion between the lever and the roller is by the ruby pin H, standing at right angles to the roller, and acting on the fork of the lever I. By the action of the wheel on the pallets, and through the pallets on the lever, the lever causes the ruby pin to move through an arc of a circle of 35 deg. on each side of the line of centres.

The small engraving on the previous page is a representation of the back of Mr. Jones's Exhibition "Rose Watch," showing Time and its doings. On the back is a Maltese cross, and on its four limbs are depicted the four seasons of life, in the bud, blossom, decay, and death of a rose.

Mr. Jones's case in the Exhibition excited great attention from the fact of the watches being placed on a revolving stand. It contained, besides the Rose Watch, various specimens of English lever watches; one of which was engraven on the back with cubical crystals—a most admirable idea for such a purpose, seeing that the primary crystal of gold is true cube. The same exhibitor also shows a new and simple mode of producing dead seconds.

We have mentioned a few of the more prominent English exhibitors of clocks and watches, reserving a notice of foreign contributors to a future occasion. We here give an engraving of DENT's TURRET CLOCK, of which we have already spoken—ante, p. 14—at some length. Mr. Dent has obtained the Council Medal for this clock on account of the combination of strength and accuracy of time-keeping attained in it, which requisites are accomplished by a cheaper mode of construction than in other turret-clocks of high character.

Mr. Dent is also a pretty large exhibitor of watches and philosophical instruments, to the consideration of which we may possibly return; though any commendation of "Dent's watches" is deemed almost supererogatory now-a-days.

While on the subject of clocks, we cannot forbear glancing *en passant* on a new design for a clock-face by Mr. Frances, of Devonshire-place. It is attempted to make the familiar face of a clock a medium through which a palpable waste of time may be more vividly brought to the remembrance of the time-wasters. Around the face of the instrument are drawn two circles, in which are inscribed numerous texts from Scripture, all tending to instruct us how "to walk wisely in a perfect way." The idea is extremely ingenious.

TURRET CLOCK.—EXHIBITED BY MR. DENT.

FIG 7.3 Dent's clock in the Great Exhibition performed very well.

timekeeping of this clock, since it had no second dial. The attempt backfired, since he pointed out it how easy it was to observe the exact instant when the remontoire was released.

In the fourth category, we see Dent's flatbed style developing from Dent's Great Exhibition clock, which was almost certainly his first, made in 1853 for Fredericton cathedral in New Brunswick, Canada. Virtually the same style and layout was used in clocks at Cranbrook, in Kent (1856), Balmoral Castle (1857), West Ham (1857), Doncaster (1858), Oakham (1858), and Leeds

Town Hall (1859). West Quantockhead also has a Dent movement, although it is neither signed nor dated, and there is another at Ramsgate Harbour. It is this group of clocks that gives us some clues as to the development of the Great Clock's escapement, since all these clocks, except the one at Ramsgate, are fitted with three-legged gravity escapements.

1853

The Death of E.J. Dent

Edward John Dent died on 8 March 1853; he was 60 years old. In his will, Dent left the turret clock factory to his stepson Frederick Rippon; desiring that he took on the surname Dent, and advising that Frederick took an able man of business into partnership with him.

Frederick Rippon

Dent had married a widow, Elizabeth Rippon, in 1843; she had had two sons by Richard Rippon who was a watchmaker. Her son, Frederick William Rippon, was born in 1808 and it was not surprising that he entered the trade; he was apprenticed to Vulliamy. By 1832, it seems that Frederick was working for Arnold and Dent and he appears to have been an excellent workman, since he submitted two chronometers to the Greenwich Trials.

The Situation in March 1853

With E.J. Dent's passing, the Great Clock was partially complete; the going train was running and probably the quarter and hour trains were well advanced. Fredricton cathedral clock was probably being built at that time as well. There can be little doubt that Denison was already used

to working with Frederick, so the changeover of the factory's ownership would have gone smoothly as far as the Great Clock was concerned. Denison appears to have got on well with Frederick, since a book authored by Frederick Dent appeared in 1855. It was entitled *Treatise on Clock and Watch Work with an Appendix on the Dipleidoscope*. This was a reprint of Denison's version of *A Rudimentary Treatise on Clocks, Watches, and Bells*, as it appeared in the *Encyclopaedia Britannica*. The dipleidoscope was an instrument designed by Frederick, which used an optical prism to determine the instant of noon accurately from the sun.

1854

In January, Benjamin Lewis Vulliamy died; he was 73 years old. His business was split up, with Frodsham taking on the turret clock side. Had Vulliamy won the order, he would not have seen his clock completed. Airy's dismissing of Vulliamy's design as *a village clock of very superior character* was a rather cruel blow. The clock would have been a very superior clock but certainly would not have been sufficiently accurate or powerful enough to strike the bells.

Vulliamy would no doubt have been delighted, had he lived another five or six years, to follow all the various disasters that befell the two Great Bells and to hear the bickering that was to take place over the clock hands. It would have been interesting to hear Vulliamy's comments on the Great Clock, had he seen it in Dent's workshop.

The Great Clock was completed in 1854. No doubt the clock had been set up on a temporary frame and then the main frame made when all the sizes and positions finalized. The two long frame members were cast with the inscription of

THE GREAT CLOCK AT WESTMINSTER.

FIG 7.4 Great Clock on test in Dent's workshop, from *The Illustrated London News*, February 1857.

maker, designer, and date of completion running along the complete length.

John Whitehurst, the other competitor, died on 21 September 1855, so, like Vulliamy, would never have lived to see his masterpiece installed and working in the Clock Tower.

In October 1856, an article about the Great Clock appeared in The Engineer. It showed an engraving that was taken from a photograph by Freeman de la Motte. Somewhat surprisingly, the clock has a remontoire fly; the text explains that the remontoire is installed so that the external hands have a discernable motion. The clock also has a gravity escapement. The engraving was reproduced in February 1857 in *The Illustrated London News*.

1857

A series of letters were exchanged in *The Mechanics' Magazine* between the chronometer maker, E.T. Loseby, and Denison. Loseby complained about the size of the bell hammer and the use of cast-iron wheels. Denison replied with his usual methodical and powerful arguments. At one stage, Loseby alluded to Denison with all the colour that Victorian abuse could muster:

Natural history tells of an animal called the squash [presumably what we today call the skunk], which has a method of fighting so offensive, repulsive, and inglorious, that most persons out of regard for their reputation and the taint which that creature leaves behind, feel somewhat ashamed of being in a contest with one of them.

After this, Denison left the correspondence, with the retort that 'English clockmakers (with very few exceptions) are the most ignorant, unprogressive, and obstinate of all the mechanical trades.' He thanked Loseby for 'giving him the opportunity of telling the public a few things that they were not likely to know otherwise.' In his next letter, Denison bowed out of the arguments. Later in 1859 Loseby was to take up the battle with Denison and to monitor the perform-ance of the Great Clock.

B. & J. Moore, of the Clerkenwell company of turret and domestic clockmakers, wrote a letter in March in *The Mechanics Magazine* to say that they could make a suitable clock for Westmin-ster for the same price that Dent was charging and that they would use gunmetal wheels. They must have known that the Great Clock was long since complete, so they knew they did not have to fulfil their promise. They also claimed that Mr James, the engineer who worked on the Houses of Parliament, had been called in by Denison to re-model the clock. Quite probably, James was involved when the issue was raised about fitting the clock into the tower.

CHAPTER EIGHT
THE BELLS: 1856 TO 1858

THE BELLS

Most British turret clocks up to the mid-nineteenth century simply struck the hours, but some sounded every quarter of an hour, using two bells to give ding-dong (or ting tang). Clocks *strike* the quarters, though the term *chiming* the quarters is loosely used. A very small number of clocks used three or more bells for quarter striking. In the eighteenth century, some churches had a tune barrel or chime; these are sometimes called a carillon. Such chimes were like a giant weight-driven music box that rang all the bells in the tower. The number of bells was mostly eight, but there could be as few as five or as many as twelve. Several times a day, after the hour had been struck, the chime would burst into action and play a tune. Apart from tunes for psalms or hymns, popular tunes of the day were often used. In the main, a chime had several tunes that were changed either every day or every time the chime operated.

It is against this background that an appropriate means of striking for the Great Clock would have been required, but oddly enough little was said or discussed about the exact nature of the hour bell or quarter striking. Vulliamy had mentioned ten bells in the tower in his proposed clock but the matter seemed to have escaped the eyes, or the ears, of the Commissioners and the public.

Edmund Beckett Denison started bell ringing on church bells at the age of 17 when he was studying with the Rev. Hodgson at Helions Bumpstead in Suffolk. At Cambridge, he rang at the university church of St Mary the Great, so he was very familiar with the clock chimes there. It was in 1793 that St Mary's needed a tune for the quarters of their new clock and the Rev. Dr Joseph Jowett composed the chime with the assistance of a pupil of his called William Crotch.

In the 1877 edition of *Encyclopaedia Britannica* Edmund wrote:

The repetition of four ding-dongs can give no musical pleasure. The case is different with the Cambridge and Westminster quarter chimes on four bells, and the chime at the hour is the most complete and pleasing of all. It is singular that these beautiful chimes (which are partly attributable to Handel) had been heard by thousands of men scattered all over England for 70 years before anyone thought of copying them, but since they were introduced by Sir E. Beckett in the Great Westminster Clock, on a much larger scale and with a slight difference in the intervals, they have been copied very extensively, and are already almost as numerous as the old-fashioned ding-dong quarters.

The chime was dubbed 'Jowett's Jig' by the undergraduates and is said to be based on a bar from the aria 'I know that my Redeemer liveth' from Handel's *Messiah*. Strictly speaking, we should call the Westminster chime the *Cambridge Quarters*, but the name *Westminster Quarters* has stuck. Edmund was later involved in new bells for Doncaster parish church and he started to gather a reputation as a bell expert. Indeed, his first book included a chapter on bells.

George Biddell Airy also graduated at Cambridge; he too would have been equally familiar with the Great St Mary's chime. Airy was a referee for the new clock that was installed in the Royal Exchange at London in 1844. The clock was made by E.J. Dent and also incorporated the Cambridge Quarters; most probably at Airy's suggestion. It was the first clock in London to sound the Cambridge quarters, although the sequence was slightly different from the one that we hear today at Westminster.

So it is highly likely that Airy and Denison did manage to agree on one thing at least, and that was that the Great Clock would sound the chimes as heard from the collegiate church of their Cambridge Alma Mater. There seems to have been no discussion or dissent either with the Commissioners of the time.

1852: THE CLOCK TOWER GROWS

Whilst the clock tower slowly advanced in height, little thought was paid to the bells for the clock. By February 1856, an engraving of the time showed the top of the tower with a complete dial (although an artist's impression) and scaffolding round the lantern. By 3 August 1857, 95½oz of gold leaf had been used in decorating the outside of the clock tower, the cost of the gold being £890.

BELL FOUNDERS IN 1854

Mears Bell Foundry at Whitechapel

In 1574, Robert Mot had a church bell foundry in Whitechapel. From this early time there has been an unbroken succession of founders in Whitechapel. The Mears family were founders from 1781 through to 1865 when Robert Stainbank took over the foundry from George Mears. Famous bells cast include the clock bell of St Paul's cathedral in 1709 and the Liberty bell in 1752. Mears cast a set of bells for the Royal Exchange in 1844, so would have worked extensively with Dent on this project. Today, the company is still on the same site, founding bells and refurbishing peals of bells.

John Warner of Cripplegate

John Warner started a bell foundry at Jewin Crescent, Cripplegate, around 1788 and died in 1820. The business was, at first, mainly involved in brass founding and then cast bells as a sideline. After John's death the business was carried on by his nephew, John, and, by 1850, the foundry was starting to get a good name for bell casting, becoming the predominant bell founder in the 1870s and 1880s. Dent used Warner bells with several of the clocks he supplied. Charles Warner succeeded John in 1852 and won a patent for casting bells. Warner bells of this period bear the Royal Coat of Arms and the word *Patent*. Warners continued founding bells after the end of World War I in 1918, but only on a small scale, the bell founding ending in 1922.

John Taylor of Loughborough

The Taylor family started founding bells in 1784; they were successors to Eayre of Kettering and dated back to Edward Arnold in 1715. John Taylor, who also made church clocks with his brother William, moved to Loughborough from Oxford in 1839. John established a bell foundry to recast the bells of Loughborough parish church. Taylors cast many bells and by the 1850s was a very up and coming business. John died in 1858 but Denison became very involved with the company, to the extent that in 1864 one of the sons was named Edmund Denison Taylor. Taylors cast the largest bell in Britain in 1881; it weighed almost 17 tons and it now hangs in St Paul's cathedral. Taylors still cast bells today, on the same site in Loughborough established by John in 1839.

LARGE BELLS CAST IN ENGLAND BEFORE 1854

The eventual requirement that the Great Clock should strike on an hour bell of around 14 tons meant that it was to be the largest bell ever cast in England. *Great Tom* at Oxford was cast in 1680 and weighed almost 6¼ tons; *Great Tom* of Westminster in St Paul's cathedral weighed just over 5 tons and was cast in 1716. It was not until the second quarter of the nineteenth century that casting large bells recommenced. Mears cast *Great Tom* (almost 5½ tons) for Lincoln cathedral in 1835 and *Great Peter* (10¾ tons) for York Minster in 1845. The big bell at York was, however, generally considered a huge disappointment. These were followed in 1847 by a bell of around 11½ tons for Montreal. In comparison, neither Taylor nor Warner had attempted to make any large bells.

Not only had Mears produced an unsatisfactory bell for York, they had had great problems at the Royal Exchange. As well as a clock, Dent supplied a carillon and Mears supplied 15 bells in 1844 and a replacement set in 1845. In 1852, Taylor provided a complete new set of bells. With this bad run of luck for Mears, Denison was probably very cautious about entrusting them with the job of casting the Westminster bells.

SPECIFICATION

In April 1852, Denison wrote to Lord John Manners concerning the bells, suggesting that the large bell be cast at the Woolwich foundry under the supervision of a bell founder. Denison was concerned that the art of bell founding was in a poor state and must have been worried that no founder was capable of casting such a large bell. He wrote again in March 1854, mentioning that a significant invention had been made in bell founding by Messrs Warner of Cripplegate.

Dent contacted the Commissioners to say that Denison had informed him that the great bell should be 14 or 15 tons and that the quarter bells be arranged as though they were the 1st, 2nd, 3rd and 6th bells of a peal of which the hour bell is the 10th. Dent added that Denison had told him that Messrs Warner of Cripplegate had cast some large bells within the last two years and that they should be included in any competition.

Denison's and Dent's letters must have started something, because in October 1854 the Commissioners wrote to Barry asking for advice on providing bells for the clock tower. At that time, Dent had finished the clock and the tower was growing. Barry replied a month later and suggested some bell founders; he also proposed that Edmund Beckett Denison and the Rev. Taylor be referees for the supply of the bells. Could it be that Barry was pleased to pass the potential problem of the bells onto someone with whom he had no natural affinity?

1855

In January 1855, Barry had to prompt the Commissioners into action, saying the masonry of the clock tower was almost up to the roof. Presumably the dial openings had been almost finished, as shown in an engraving in *The Illustrated London News* the following month. So the First Commissioner, William Molesworth, finally wrote to Denison, requesting him to provide a specification for the bells and to be a referee.

Denison replied the following day in his usual forthright manner proposing:

That unless the quotation of Messrs Warner should be much higher than the others, it will, in my opinion, be desirable to employ them, because they have introduced a new patented method of casting bells, and have within the last few years been making experiments from the suggestions of the Rev. Mr. Taylor and myself, in order to arrive at the best form and thickness for large bells. And (as I stated in my letter to the Commissioners in 1852) all the very large bells which have been cast in this century have been very inferior to what I should require before I gave my approval of them for the Clock Tower at Westminster.

Mr Taylor, whom I propose as joint referee with myself, is the gentleman referred to in that same letter of mine as the person who knows more about bells than any person I am acquainted with; and, from conversations with him, I have no doubt he will be ready to give the Commissioners his assistance with me.

PROPOSED FORM of SPECIFICATION *to be sent to Messrs Mears, Whitechapel; Messrs Warner, Jewin Crescent, Cripplegate; and Messrs Taylor, Bell-founders, Loughborough.*

Gentlemen,

THE *Commissioners of Her Majesty's Works, &c., are ready to receive from you a tender for supplying the five bells for the great clock of the Houses of Parliament.*

The fifth or hour bell is to weigh not less than 14 nor more than 15 tons; and the quarter bells are to be of such notes that they would be respectively the first, second, third, and sixth of a peal of ten, the hour bell being the tenth.

They are to be fitted to stocks of such size as may be approved by the referees, but they will not be required to swing, and therefore are not to have gudgeons. They are to be furnished with clappers, which may be rung with a rope on

occasions, if required. As it is understood not to be possible to calculate exactly the weight of the bells before they are cast, the tender is to state at how much per cwt you will supply them, and also your estimate of the probable weight, and, therefore, the probable cost of the peal, assuming the great bell to weigh 14 tons, with a separate item for the cost of the fittings, and delivery at the Clock Tower of the Houses of Parliament, and the time at which you will engage to deliver them.

The whole of the work is to be done, subject to the directions and approval of Edmund Beckett Denison, Esq., one of Her Majesty's Counsel, and the Rev. William Taylor, FSA, of 73, Oxford-Terrace; and payment is not to be made until they have certified their final approval of the bells after they are hung in the tower; the Board of Works taking the expense and risk of raising and hanging them, and of any damage which, in the opinion of the referees, may have occurred to the bells after they have been delivered.

Tenders to be sent in (say in about a week after the letters are issued).

3 February 1855   (signed) E.B. Denison.

The Commissioners duly answered five days later to Denison, saying:

In reply, I am directed to inform you, that the Board approve of the specification with which you have been good enough to furnish them, with the exception that they think that the name of the Chief Commissioner should be added to those of yourself and Mr Taylor as the referees, and also, that the condition "that the Board shall take the expense and risk of raising and hanging the bells, and of any damage which in the opinion of the referees may have occurred to the bells after they have been delivered," should be omitted.

Denison went back with his usual promptness and blunt directness:

If I had any reason to believe that the Chief Commissioner of Works is conversant with the art of bell founding, I should be very glad to have his assistance, together with that of Mr Taylor, in determining the form and constitution of the bells for the Westminster clock. But assuming (for the sake of argument) that the individual, Sir W. Molesworth, is conversant with bell founding, I should, in that case, require that he should be the third referee nominatim, and not the Commissioner for the time being, as it is extremely unlikely *that two Commissioners in succession will possess the requisite acquaintance with a subject to which very few persons indeed have paid any attention, not even those who have written books or tracts on the history of bells. If the present Chief Commissioner is unable to assure me that he has paid special attention to this subject, and has some opinion whether (for instance) the thickness of the sound-bow of a bell should be 1-12th or 1-15th of the diameter, and what the thickness of the waist and the diameter of the crown should be, or the proportion of the height to the diameter, then I beg to decline acting as a referee with a person who must be incompetent to give any useful directions on the subject of the reference. These are all matters requiring great consideration, and some further experiments, which must be conducted under the superintendence of the referees; and I will be no party to any such absurdity as that of subjecting the bell founder to the control and the competent referees to the interference of an incompetent one, who is, moreover, pretty sure to act under the advice of somebody else behind the scenes. For the same reason, the Astronomer Royal and I refused to allow the interpolation of any more referees as to the great clock, when it was attempted by Sir W. Molesworth's predecessor, who, it turned out, was acting on the suggestion of some disappointed candidates for the job; and I have reason to know that you will find Mr Taylor equally determined with myself not to act with any person whom we do not know to be competent to give useful directions to the bell founders.*

I have no objection, if the Board prefers it, to the founder being required to hang the bells as well as to deliver them. I only put it as I did, because there will probably be hoisting machinery in the tower belonging to the builder, of which the Board can have the use at a small cost, but which the bell founders cannot reckon on having, and they will therefore increase their tenders on that account; and also because it may happen that the bells are delivered some time before it is convenient to hang them, and in that case they obviously must remain at the risk of the Board, unless the bell founder is to have a further sum as a kind of insurance against risk of damage. But perhaps, instead of requiring the founders to state their own time for supplying the bells, it will be better to ask Sir C. Barry when the tower is likely to be ready for them (reminding him that the great bell cannot be taken up inside, on account of the internal walls), and then require the founders to complete and deliver the bells at that time, if it is not too soon for them.

BECKETT DISMISSED

Six days later, on 15 February 1855, the Commissioners secretary replied with a dryness that matched Denison's robustness:

I am directed by the Commissioners to inform you, that, under the circumstances you have therein stated, they will not give you any further trouble in the matter.

Thus Edmund Beckett Denison was barred from proceeding with the bells. Perhaps he smiled a wry grin, knowing that one day they would probably return asking again for his assistance.

Charles Wheatstone was then contacted by the Commissioners, requesting him to act as a referee and to be in touch with Charles Barry. Wheatstone replied in the affirmative, saying that he and Barry were going to Paris to be Jurors on the Industrial exhibition and that they could examine chimes whilst they were in France. The Commissioners agreed in June and there the matter slept until after Sir William Molesworth died in October of that year, 1855.

BENJAMIN HALL... 'BIG BEN'

Benjamin Hall was born in 1802, the son of an industrialist. He entered Parliament as MP for Monmouth in 1832 and held the seat for five years. He campaigned against any abuse of privilege and was a strong supporter of the Welsh language. He became Sir Benjamin Hall in 1838 and succeeded Sir William Molesworth as Chief Commissioner of Works. He became a peer as Baron Llanover in 1859 and died in 1867. Since he was a man of very large stature he gained the nickname *Big Ben*. Sir Benjamin was buried in Llanover churchyard in an impressive tomb that he had designed himself.

FIG 8.1 Sir Benjamin Hall, First Commissioner of Works in 1856, after whom the Great Bell was named.

FIG 8.2 Sir Benjamin Hall's signature.

DENISON RETURNS

Sir Benjamin Hall, First Commissioner of Works was the successor to William Molesworth. Records are not available at this point, but it would seem that Denison was again asked to act as referee for procuring bells for the Clock Tower.

1856: WARNER CASTS THE GREAT BELL

Messrs Warner & Sons of Cripplegate won the contract to supply the bells; their patent system had appealed to Denison. Their patent involved

METHOD OF MAKING THE BELL-MOULD.

FIG 8.3 Preparing the core and cope for casting the bell at Norton on Tees, from *The Illustrated London News*, 25 August 1856.

using a large cast-iron container that was generally the shape of the bell but somewhat larger. This container, called a cope, supported the moulding material when the bell was cast, thus being a more reliable way of preserving the shape of the mould rather than the bands of iron used previously. Warner's Crescent foundry at Jewin Street, in Cripplegate in London, did not have a large enough furnace to cast the great bell, so it was decided to use a new foundry that one of the family had established at the village of Norton close to Stockton-on-Tees.

Two large furnaces, each capable of melting 10 tons of metal, were specially built at the Norton site of Messrs Warner, Lucas, and Barrett. It took six weeks to prepare the mould for the Great Bells. A core of bricks was built in a brick-lined pit in the foundry floor; this core was then covered with the traditional bell-founders mixture of loam, horse hair, clay, sand, and horse manure, and smoothed to the exact profile with a shaped piece of wood called a strickle or sweep that turned round a central post. The core set the internal shape of the bell; the outside shape was determined by an outer mould. This time, the loam mixture was applied to an up-turned cast iron cope and a different strickle employed to give the outside profile. Inscriptions and decorations were impressed in the outside mould to give a raised impression on the bell. When the moulds had been thoroughly dried, they were securely fitted together and buried in the pit with well-rammed sand and iron pigs set round the outside. A channel was dug in the foundry floor to carry the molten bell metal from the furnaces to the open mould.

On the evening of 5 August 1856, the furnaces were fired and fuelled until they had reached the correct temperature. In the early morning of the 6th, an hour was taken to load the metals, and two and a half hours later the whole 18 tons had become totally fluid. On a signal, the furnaces were tapped and the metal ran into a holding pool; there the flow was controlled by an iron sluice gate. In only five minutes, the mould of the Great Bell had been filled and the collected watchers joined the workers in giving three hearty cheers.

THE CASTING OF THE BELL FOR THE GREAT CLOCK OF WESTMINSTER PALACE, AT NORTON, STOCKTON-ON-TEES.

FIG **8.4** Casting a bell is always a memorable event, as demonstrated by the worthies watching. What the engraving gives no feeling for is the incredible heat from the river of metal and the smell of smoke and heated moulding material. From *The Illustrated London News*, 18 September 1856.

Two weeks later, on 22 August, the bell was dug out of the pit and was sounded for the first time. The bell stood 7 ft 10½ in high and 9 ft 5½ in in diameter and its note was E natural. The inscription ran *Cast in the 20th year of the reign of Her Majesty Queen Victoria and in the year of Our Lord 1856 from the design of Edmund Beckett Denison QC; Sir Benjamin Hall, Baronet, MP, Commissioner of Works.* The Royal Coat of Arms and Warner's name also appeared on the bell. The bell weighed 15 tons, 18 cwt 1 qrs 22 lbs.

The Illustrated London News reported that the alloy was made of 7 parts of tin and 22 parts copper and was harder than normal bells; nearly as hard as spring steel and harder than modern bells. In the first week of September, the bell was on public view, for an entry fee of 6 d., or 3 d. for a child, visitors could see the monster bell. Over three days, £8 was collected and was given to Norton Literary Institute.

From Norton, the bell was taken by rail to West Hartlepool dock to be shipped to London. Sunday, 7 September was chosen for the journey, since the wide load would have interfered with other trains on the line during weekdays.

The vessel *Wave* was chosen to carry the Great Bell to London from West Hartlepool; Mr Moncrief of Wisbech was the master. *Wave* was a Billy-Boy schooner, a sailing barge of 105 tons, ideal for coastal and river work; the design originated from boats on the Tyne and Humber. Such a craft had one or two masts and could have been up to 75 ft long with a beam of 18 ft, having a flat bottom that made it convenient for work in shallow water. At sea, bulwarks provided protection and since it had no keel of any depth, lee boards were used when sailing—these were large wooden boards that were lowered over the side of the craft to work as keels. Additionally, *Wave* could lower her masts to allow passage

FIG **8.5** Having been broken out of its mould, Big Ben I is seen to be a giant bell. Warners routinely added the Royal Coat of Arms and the word '*PATENT*' to bells they cast.

THE QUARTER-BELLS FOR THE GREAT CLOCK AT WESTMINSTER.

FIG **8.6** The quarter bells were also cast by Warners; the largest was cast at Norton, the others in the London foundry at Cripplegate.

BIG BEN OF WESTMINSTER.

THE BELL CAST BY
MESSRS. WARNER & SONS, FOR THE
CLOCK TOWER OF THE NEW HOUSES OF PARLIAMENT.

DIMENSIONS.—*Diameter at Mouth 9ft. 5½in.—Height 7ft. 10½in.*
WEIGHT.—*With Clapper, 16 Tons 11 cwt. 2 qrs. 20 lbs.*

The casting of this monster bell was accomplished on the morning of the 25th of August, 1856, by Messrs. Warner and Sons, of London, at the furnaces of Messrs. Warner, Lucas, and Barnet, situated in the picturesque village of Norton, near Stockton-on-Tees.

FIG **8.7** This card was probably produced by Warners as a souvenir of the casting of Big Ben I. The image is a copy of the bell as it appeared in *The Illustrated London News*.

97

under river bridges; essential, since the plan was to unload the bell at Lambeth. In June 1954, an auxiliary barge called *Wave* was photographed at moorings at Newport, Isle of Wight. It was described by the National Maritime Museum as '*Over 100 years old, still trading.*' This boat was registered at Cowes, Isle of Wight, in 1912 but the place and date of build were not given in the Merchantile Navy List. This craft must have been the same *Wave* that carried the Great Bell.

Shear legs were erected at West Dock for the bell but they were too small and a larger pair was erected at Swainson Dock. At 1 p.m. on Saturday, 13 September, work started on loading the bell. It was swung safely over the hold, where it was discovered the hold was not wide enough for the

FIG **8.8** *Wave*, the Billy-Boy Schooner, carried Big Ben I from Hartlepool to London. The National Maritime photo index records '*An auxiliary barge, ex-schooner, at moorings Newport Isle of Wight, over 100 years old. 10 June 1954.*' Photo courtesy of the National Maritime Museum.

bell to pass through. Another deck plank had to be cut out of the vessel. When all was ready, one of the chains had slipped down one side of the sheave and the blocks refused to work. After a delay of an hour, all was again ready; a slight crack had been heard earlier but ignored. As soon as the strain was put on the shears, they swerved to one side and fell to the ground with a great crash. The foremast was brought down, seriously damaging the hull. The bell fell to the bottom of the boat, the vessel started to take in water, and, to prevent her sinking, two steam-boats took her in tow and removed her into the outer basin where she was run aground on a sandbank, it luckily being high tide. The Captain of the vessel was in the hold when the accident happened, but he and the large crowd who had gathered were unharmed. The whole incident was witnessed by one *J.H.*, who also took two calotypes of the accident.

The *Stockton and Hartlepool Mercury* for 20th Sept 1856 reported that *Wave* spent a time in Mr Pile's graving Dock. The bell was removed from the hold and suspended by means of blocks and strong chains to two baulks of timber fixed horizontally across the hold. It was held several feet in the air to clear it from the deck. *Wave* eventually sailed for London on the evening of Sunday the 5th. An insurance of £3,000 had been taken out on the bell in transit.

THE NAME

Although one report stated that the *Wave* was nearly lost in a storm, the Hartlepool local paper reported that on Tuesday morning the boat safely delivered her cargo alongside Messrs Maudsley's wharf near Westminster Bridge. The newspaper reported that in compliment to the Home

Secretary the Westminster Great Bell cast at Norton was to be called *Big Ben*.

In reports in *The Times* reference was always made to the Great Bell until 22 October 1856, when the newspaper announced that the bell would be named '"*Big Ben*" in honour of Sir Benjamin Hall, First Commissioner of Works.' Sir Benjamin had been in office for 18 months and had restarted the process for procuring the bells that his predecessor Sir William Molesworth had not completed. As such, Sir Benjamin, the sort of man who got on with the job, as well as First Commissioner for Works, had his name enshrined in our national icon. There is an apocryphal story that was recounted in a letter to *The Times* in 1959 by Edmund Esdaile, whose great-great-grandfather was a cousin to Sir Benjamin Hall. He told how the House of Commons debated the name for the bell and when someone called out 'Call it Big Ben' (Big Ben was a nickname for Sir Benjamin) the name received applause and universal approval. Mr R. Gresham Cooke, from the House of Commons, soon researched *Hansard* and found no record of the debate; he suggested that such a source for the name may have come from a standing committee on the Parliament buildings or from the smoking room. Esdaile did some further research with family members and wrote back to report that the name was arrived at as described, but not in a debate; probably in the library or smoking room. Miss Maxwell Frazer, a biographer of Sir Benjamin Hall, wrote a letter to *The Times* in March 1971, in which she described an interview with Major John Berrington, whose father was Sir Benjamin's private secretary. John Berrington confirmed that the name 'Big Ben' had been decided in a committee. The term 'Big Ben' at that time in the nineteenth century appeared to be a current appellation that could be applied to

anything of giant size; the bell was big and so was the First Commissioner. It was likely that Parliament had a sense of humour and had their way with the name.

There is a suggestion that the bell's name came from *Ben Caunt*, a famous prize fighter of the day, who stood over 6 ft 2 in tall. *Punch* made the proposal in a mocking article that must have been aimed at the Government naming the bell instead of seeking public consultation. From this start the myth was perpetuated.

Perhaps *The Times* just named the bell and so it stuck and became *Big Ben*. Other suggestions mentioned in subsequent newspaper letters were *Victoria*, after the Queen, *Albert*, after the Price Consort, or *Great Edward*, after Edward the Confessor. The satirical magazine *Punch* was soon on the scene, suggesting that the clapper be called *Gladstone*, after the person in Parliament who had the loudest tongue.

ARRIVAL OF THE BELL

By 10 October, *Wave* had not arrived in London; this must have given some concern for its safety. However, it transpired that she did not leave until the 9th. It was reported that the bell was unloaded on the 21 October, so it certainly looked as though the voyage was a difficult one.

On arrival at Maudsley's (the famous engineer and designer) wharf at Lambeth the Great Bell was lifted with a massive crane without mishap. From Lambeth, a team of 16 horses pulled the bell on its carriage across Westminster Bridge to New Palace Yard, where it was received by Mr Quarm, the Clerk of works, Professor Taylor, and Sir Charles Barry, the architect. Masses of onlookers turned out to see the bell and the police had considerable difficulty keeping the

FIG **8.9** Unloading Big Ben at Maudsley's wharf at Lambeth, from *The Illustrated London News*.

THE UNSHIPPING OF "BIG BEN" AT MESSRS. MAUDSLAY AND FIELD'S WHARF, LAMBETH.

approaches to the Palace clear. Here at the base of the clock tower a construction of massive timbers had been built to hang the bell; during the afternoon it was lifted off the carriage, suspended from chains, and given some blows to hear it speak. The tone was pronounced as good, and the bell propped up with wood to relieve the chains.

FIG **8.10** Sixteen horses pulled the bell triumphantly from Maudsley's wharf to New Place Yard, from *The Illustrated London News*.

REMOVAL OF THE BELL FOR THE GREAT CLOCK OF WESTMINSTER PALACE.

FIG 8.11 An enormous clapper was made to go inside the bell; it weighted 13 cwt.

THE CLAPPER

A large wrought-iron clapper for the Great Bell was forged by George Hopper of the ironworks at Houghton-le-Spring near Durham. The clapper was then smoothed and finished at his new works near Fencehouses. It weighed in at 1485½ lb, and the ball at the end was a huge 24 inches in diameter. The intention was to toll the bell for important national events. In an engraving in *The Illustrated Times*, (22 November 1856) the clapper can be seen in use to test the bell when it was hung in Palace Yard. Most probably, the first clapper was scrapped when the bell had to be recast. Big Ben II had a clapper as well; it weighed 6 cwt but was never used and was most likely removed from the bell in 1956.

TESTING THE BELL

Once the bell was securely hung from the stout timber frame, it could be tested; this seems to have occurred in two stages. The *Illustrated Times* reported that there was a private sounding for those involved on Thursday, 13 November 1856, just after the clocks of St Margaret's, Westminster Abbey, and St Paul's had struck 11 a.m., a team of men led by Edmund Beckett Denison pulled on a rope attached to the clapper. The great bell duly boomed out. *The Illustrated London News* of 27 December reported how a hammer of almost half a ton was raised by a crab and then released; the experiment being repeated for different heights of hammer fall. On Saturday, 31 January 1857, the bell was rung for an extended period, Denison was demonstrating the bell to some campanologists; not, as some people thought, announcing a Royal birth. Two days later, the bell was tolled for the opening of Parliament. The Times also pointed out the fallacy of not casting the bell before the clock was made.

W.L. BAKER COMPLAINS

In 1854, W.L. Baker was awarded a patent for hanging bells; this involved suspending the bell from a central bolt, which enabled the bell to be turned easily when the clapper caused wear on

FIG 8.12 W.S. Sargent produced this ink and wash drawing, dated 22 October 1856. It shows Big Ben in the process of being raised from it wheeled trolley. Courtesy of the London Metropolitan Archives.

the soundbow. Baker claimed that Denison had appropriated his patented principle for the means for hanging the Great Bell, i.e., it had a button-shaped top secured by a large collar that would enable the bell to be turned. Denison's idea was not to allow for wear, but that the bell could suspended in a manner that allowed it freedom to vibrate fully. Denison responded, in his book on building, in true legalistic manner by saying that you could not patent a principle, only an

EXPERIMENT WITH THE HAMMER UPON THE GREAT BELL FOR THE WESTMINSTER CLOCK.—(SEE NEXT PAGE.)

FIG **8.13** Once Ben had been raised on a scaffold, he was sounded every day for the benefit of the public, from *The Illustrated London News.*

implementation of a principle. Baker felt sufficiently aggrieved to compose an open letter to Denison, it was written from Glasgow, dated 1 January 1857, and published as a pamphlet for the price of 4 d.

1857

Big Ben Cracked

Big Ben continued to be sounded for the entertainment of the public, but late in October 1857 it cracked. The crack, which was directly opposite where the bell was struck with the 12 cwt hammer, started at the lip, went through the soundbow and extended 40 inches up the waist of the bell. The fracture appeared clean, but the metal had a coarse dull appearance, and was full of very minute holes, which were noted on other parts of the bell when it was afterwards broken

up. Further examination showed that Big Ben was much thicker in the waist than Denison had intended, which, he explained, was why the bell was heavier than expected, was half a tone lower, and needed such a heavy hammer to bring out the full note. It is likely that the cope of the bell 'floated up' a fraction due to the immense weight of the bell metal inside, and even though the earth had been well rammed around the bell mould, some movement probably took place, resulting in the extra separation of the inside and outside moulds and thus the increased thickness of the bell.

An engineer, Charles May, reported that he had been present at the experiments and had heard the bell crack very distinctly; he mentioned it at the time that the blow was struck and the fact was soon afterwards verified. Other reports said that, *'The huge external hammer was not always recovered by the recoiling apparatus and it chattered on* the bell.' A hammer chattering on a bell, certainly a very large one as was used in the instance, is certain to damage a bell.

Denison and Taylor inspected the bell on 24 October and condemned it. It was soon decided by Denison, Professor Wheatstone, and First Commissioner Benjamin Hall that there was no alternative but to have the bell recast. Warner tendered for recasting but the job went to Mears of Whitechapel, since they were significantly cheaper; it was expected that the recasting would cost about £600. A slight change in design would reduce the diameter and thickness of the bell and thus return the note and weight to that originally intended.

Denison wrote a four-page letter to Sir Benjamin the First Commissioner on 25 November 1857, the gist of which was that Warners were at fault for the cracking of the bell. Presumably in his defence, he methodically attacked Warner on every front. Denison said that he went to Stockton to see the new bell and Warners said to him, 'You were wrong about the weights: it's 16 tons instead of 14 tons.' Denison continued, 'The waist or unmeasurable part of the bell was ¼ inch thicker than I ordered. I measured all that could be measured without a special instrument for the purpose.' He added, 'Warners had been struggling for more thickness than I was inclined to allow, though I did allow more than was used in modern bells.' He went on to say, 'It was very far from satisfactory with the change of note and you know that Mr Taylor signed the certificate reluctantly, but the bell seemed so good in itself that it would have been thought a very harsh thing to reject it as we unquestionably could have done without any further enquiry.' In conclusion, Denison said of Warner, 'They had better shut up their foundry for nobody in their senses will employ such a founder. Warners have been paid £130 for learning their business (referring to

experiment bells cast) and they richly deserve to lose their profit.'

Why Big Ben Cracked

Sir Benjamin Hall asked the opinion of Denison, whose reply was given to the House of Commons on 10 December 1857 and recorded in Hansard.

My dear Sir,—The cause of the cracking of the great bell was correctly stated in The Times of 19 November.

Through some mistake or accident, which the founders say they cannot account for, the waist (or thin part of the bell) was made one-eighth inch thicker than I designed it. The consequence was that it required a clapper of twelve cwt instead of seven cwt to bring the full sound out; and although the sound-bow or striking part of the bell bore this clapper for nearly a year, it gave way at last.

If the bell had cracked when it was first tried with this clapper the founders would undoubtedly have had to recast it at their own expense, as they had engaged to make it according to my drawings, and in fact had refused (as you know) to undertake the job unless I would take the responsibility of designing all the bells. But as it seemed to be able to bear this large clapper, even when pulled by ten men, and was generally approved, Mr Taylor and I did not feel justified in withholding our certificate on account of the deviation from the prescribed thickness, weight, and note; for the quarter-bells were not then made, and could as easily be adapted to the note E natural as to E flat, which I had intended the great bell to be. Moreover, we were quite satisfied of the soundness of the casting by various tests, and, therefore, Messrs Warner were paid, and cannot now be fixed with the cost of recasting.

It is right, however, to add that this recasting will not bring the whole cost of the bells above the original estimate, as it included a large margin for contingencies on account of the obvious risk of casting five such bells as these in tune with each other, the smallest of them being as heavy as the largest bell of an ordinary modern peal, and the largest fifteen times as heavy. There is no further risk now, as the quarter-bells are all cast, and only exceed my estimate of eight tons weight by twenty-nine pounds.

The new bell is to be delivered sound by 19 February, and is not to be paid for until it has been tried by ringing with a seven cwt clapper, nor unless it agrees with the prescribed dimensions, composition, and specific gravity. Nothing more can be done to secure its goodness and its durability.

The best bells in the world, such as the old Great Tom of Lincoln, sometimes crack quite capriciously, while very bad ones, such as Great Tom of Oxford, sometimes obstinately last a great deal too long.

The great bell of the Roman Catholic cathedral at Montreal cracked in about a year after it was hung. So in that respect, we are no worse off than other people; and, luckily, our bell was not hung before it cracked.

Yours truly,
E.B. DENISON

Big Ben is Broken Up

The Illustrated London News reported on the fateful day of Thursday, 18 February 1858 that saw the passing of the great bell.

With the Rev Taylor and Edmund Beckett Denison in attendance, like mourners at a funeral, they witnessed Big Ben being lowered from the frame and tilted on his side. Next an iron ball of 24 cwt was raised up with a block and tackle to the full 30-foot height of the frame. It paused briefly

and then was allowed to fall. It dealt a death blow and two huge fragments broke out of Ben, the bell sounding its last mournful voice in protest. It took nearly a week to reduce Ben to pieces of sufficient size that could be easily moved. These were gathered together and taken off to Mears' bell foundry in Whitechapel.

'Speckliness' was reported throughout the casting. One question was, 'Would the bell be another 'Big Ben', or would it be a 'Big John' after the new Commissioner of Work, Sir John Manners?'

Who Was to Blame?

Somehow the finger of blame, though pointed around, never settled on anyone. Denison blamed Warner for casting the bell too thick, but said that the casting was a good one and dismissed the report of speckliness. There were claims that a flaw was found where it was believed that the two streams of metal pouring into the bell met. Warners blamed Denison for specifying a bell metal alloy that was too brittle and for using an excessively heavy hammer to sound the bell.

CLOCK BELLS FOR THE NEW PALACE OF WESTMINSTER.—BREAKING UP "BIG BEN."—(SEE NEXT PAGE.)

FIG **8.14** Once cracked Big Ben 1 was pounded with a large iron wrecking weight. It took almost a week to reduce the bell to manageable pieces.

Somehow the matter was forgotten; the general view was that it was good that the bell cracked when it was on the ground rather than after it had been installed in the tower when the expense of replacing the bell would have been huge.

QUOTATIONS FOR A NEW BIG BEN

Towards the end of 1857, the Commissioners approached both Warners and the Whitechapel Bell Foundry for a price for casting the new bell. They also applied to Maudsleys of Lambeth, who though not bell founders were an engineering company used to making large castings. They all offered similar prices for the scrap bell metal and for recasting the Great Bell. Whitechapel quoted £177 6s. 8d. per ton, Warners £200 0s. 0d. per ton and Maudsley £186 13s. 4d. Clearly, Whitechapel was the lowest price and were awarded the contract. An order to add the Royal Coat of Arms, the Portcullis of Westminster, and the inscription was given to Whitechapel on 8 December. The inscription was to have blank blocks, where the weight was to be entered after the bell had been weighed.

A little ditty from *Punch* circulated in London:

Poor Mr. Warner is put in a corner
For making a bad Big Ben.
And now it appears that the good Mr. Mears
Is to furnish a new Bell. When?

NEW BIG BEN CAST

Mears used the older method of bell casting, which involved making a dummy bell of clay; the mixture of loam was exactly as Warner used; the formula that had been in use for centuries. Preparations for the bell mould had taken some five months, commencing with the building of a

FIG 8.15 Mears cast Big Ben II in the evening, after a day of melting the metal.

RECASTING THE GREAT BELL FOR THE CLOCK TOWER, NEW HOUSES OF PARLIAMENT.

pit in the foundry floor, 15 ft in diameter and 15 ft deep. The core was constructed with 1,300 fire-bricks. Iron stays were used to ensure that the core retained its shape under the enormous pressure of 14 tons of bell metal. Loam was used to cover the bricks and the inner shape of the bell thus formed. Once the core had been constructed and very thoroughly dried, clay was applied and a strickle rotated round the core to give it the correct shape. The net result was that a clay bell, a perfect model of the final item, was built up over the core like a cloak. Decorations and inscriptions were added to the clay model; sometimes these were made of wax and applied to the bell surface and melted out before the final casting was done.

Over the core, with its clay model bell, layers of loam were applied to create a cope. Once a thickness of around 12 inches was reached, the cope was secured firmly with horizontal hoops and vertical stays of iron. When dry, the cope was lifted up taking with it the clay model of the bell, this model was then broken out of the core. Ben's cope weighed 7 tons; it had to be massively strong to contain the weight of the molten metal. The cope and core were then united on an iron base, securely clamped together, and well baked before casting took place. Before the casting took place, earth was securely rammed all around the cope, filling up the casting pit to the top. Blocks of cast-iron were placed around the top to ensure that the metal did not cause the earth to float up.

Saturday 10 April 1858 witnessed the recasting of the great bell *Big Ben* the second; the name was firmly stuck to the bell by the newspapers. Edmund Beckett Denison again specified the bell's metal composition, of 7 parts of tin to 22 of copper. He also designed the shape and ordered that the mould be preheated with hot air blown through it for several days before the casting took place. It took 20 hours to melt all the metal. Mr Mears still used wood as a fuel, which was slower than using coal but thought to give a better result. At 7.33 in the evening the metal was considered ready, the hot air blast to the mould cut off and the furnaces tapped and the molten bell metal run into the mould buried in the foundry floor. It took 21 minutes to fill the bell mould. For some unknown reason, Edmund Denison was not present at the casting. Mears had to wait for two weeks for the bell to cool and to see if they been successful or not.

Once the bell had cooled, been broken out of its mould and lifted from its pit, it sounded the note E. Denison pronounced that the bell was a fine one. With Big Ben's note established, the quarter bells were to be retuned to suit the new hour bell.

In the Whitechapel bell foundry day book for 24 May 1856, the price for casting the new bell was listed as £2,401 16s. 9d. From this figure, a deduction of £1,829 14s. 9d. was made for the old bell metal that weighed 15 tons, 19 cwt 2 qrs 19 lb. On 28 May, George Mears submitted his invoice to the Commissioners for recasting of the Great Bell; it was for £572 0s. 0d.

The head founder at Mears was Thomas Kimber. To demonstrate the principle of bell-founding, Thomas made a small model bell of boxwood with its accompanying core, cope, and strickle. This is now on display in the Whitechapel Bell Foundry's museum. Thomas was also an accomplished artist and in 1866 he painted a watercolour of the Great Bell for Mr Stainbank, the new owner of the foundry. The legend on the bottom gave the details of the casting; this was done in a gilt panel and today the gold leaf is as bright as it was when it was first laid. Also in the museum are the letter stamps used for making

FIG **8.16** The left-hand strickle on the wall is the original pattern to make Big Ben. Acknowledgements to Whitechapel Bell Foundry.

FIG **8.17** The drawing of the Great Bell made by Thomas Kimber. Courtesy of the Antiquarian Horological Society archive.

the inscription and copies of the Royal Coat of Arms and the Portcullis arms of Westminster. Over the entrance door of the foundry is a full-sized wooden section of Big Ben; in the foundry itself mounted on a wall is the strickle that was used to give Ben his shape.

ARRIVAL OF THE QUARTER BELLS

In early February 1857, the first of the quarter bells, the largest, was cast at Warner's Norton foundry. It was 6 feet in diameter, weighed a little under 4 tons and its note was B. By mid-October, the other quarter bells had been cast;

their notes were reported as E, F♯, and G♯; all being an octave above the hour bell.

On 6 March 1857, Mr Denison lectured to the Royal Institution on bells; he covered the history and shape of bells and, of course, the casting of the Westminster bell. Henry Wylde was a doctor of music, who could not resist commenting in *The Times* on the Great Bell when he first heard it; and on Denison's lecture. Of course, Denison quickly replied to Wylde to put him in his place.

In the fourth edition of *A Rudimentary Treatise on Clocks, Watches, and Bells*, Denison says that the hour bell at Balmoral Castle is a half-sized version of Westminster's fourth quarter bell. Since the Balmoral bell was cast in 1856, it may have been one of the experimental bells produced while

FIG **8.18** A wooden profile of Big Ben inside the Foundry's door. The author inside is 6 feet tall. With thanks to the Whitechapel Bell Foundry. This figure is reproduced in colour in the colour plate section.

Denison was working with Warners on the sizes and shapes for the Westminster bells. Like Westminster, the Balmoral bells have button-shaped tops to allow for easy turning and that were also intended to allow the bells to sound freely.

The fourth quarter bell was reported as having a fine tone, and it was hoped that the new Big Ben should be in tune with it. Sizes and weights were given.

In the House of Commons on 7 May 1858, *Hansard* recorded Sir John Manners (The First Commissioner of Works) as saying, that in answer to the question of the Hon. Member for Peterborough (Mr Hankey), he had heard that day that the Bell for the Clock Tower had been recast, and would probably be fit for delivery in ten days. He was glad to say that there was every prospect that the new Bell would have a remarkably fine tone.

Sir William Frazer, the MP for Barnstaple, replied, that as the great bell which had been originally cast had been generally known as 'Big Ben', he wished to know the intention of the Government with regard to the nomenclature of the new bell. If the Government would allow him to throw out a suggestion, he thought the new bell would be very appropriately designated 'Little John'. No doubt Sir William's suggestion brought a peal of laughter.

ARRIVAL OF THE NEW BIG BEN

Big Ben was moved out of the foundry at Whitechapel on a low carriage loaned by Messrs Maudsley. It was drawn by 16 horses and passed over London Bridge, along Borough Road, and across Westminster Bridge, arriving at the Palace about 11 a.m. on 28 May 1858. Since the bell had not been weighed, the inscription on the bell that recorded its weight was left blank. Owing

ARRIVAL OF THE NEW BELL "VICTORIA" AT THE CLOCK TOWER, NEW PALACE OF WESTMINSTER.

FIG **8.19** Like its predecessor, Big Ben II was given a great welcome when it arrived at Westminster. The print is entitled 'The Arrival of the new Bell 'Victoria' at Westminster', from *The Illustrated London News*, 5 June 1858.

to new Ben's slightly different note, Warners had to retune the four quarter bells. A full-page engraving shows the procession passing over Westminster Bridge, the caption *The Arrival of the new Bell 'Victoria' at the Clock Tower, New Palace of Westminster* indicates that a name change for the bell was being considered.

Big Ben II was weighed on 10 June by the Weighing Clerk to Messrs Stewart & Co., using their steelyards; a certificate was issued and countersigned by Thomas Quarm. The numbers were then duly added to the bell's inscription by chiselling them out the blank sections that had been left proud. The inscription that runs around the lower part of the decorated bell reads: *This bell, weighing 13 tons 10 cwt 3 qrs 15 lbs was cast by George Mears of Whitechapel for the Clock of the Houses of Parliament under the direction of Edmund Beckett Denison QC in the twenty first year of the reign of Queen Victoria and in the year of our Lord MDCCCLVIII.* It is just possible to make out the raised parts that were left so that the figures for the weight could be added. Gothic tracery

surrounds the top of the bell. On one face of the bell is cast the Royal Coat of Arms, on the reverse the Portcullis of Westminster.

Together Denison and Taylor signed a certificate, dated June 1858; it said: 'In pursuance of the contract made with Mr Mears for the casting of the Great Bell for the Houses of Parliament, we hereby certify that the bell has been recast and delivered to the foot of the Clock Tower and that we approve of the same.'

INSTALLATION OF THE BELLS

By October 1858, the work on the Clock Tower had progressed sufficiently to enable it to receive the clock and bells. The tower had been built, the slightly projecting dials added, and the belfry, with its ironwork to support the bells, was mostly installed. The lantern was still being constructed. Whether the bell was to be called *Big Ben* or the *Royal Victoria* was still undecided.

Bells cast by Messrs Mears.	Diameter of Mouth Feet. In^s	Thickness of ound Bow Inches	Weight. Tons Cwt. Qrs. lbs
Westminster	8 . 11⅞	8¾	13 . 10 . 3 . 15
Montreal	8 . 7½	7⅞	11 . 11 . 1 . 0
York	8 . 3½	7¾	10 . 15 . 0 . 0
Lincoln	6 . 10½	6	5 . 8 . 0 . 0
Canterbury	5 . 9	5½	3 . 10 . 0 . 0

Messrs Mears were first Established in 1738, & have cast 571 Peals, containing 3811 Bells & Single Bells, from 4 Cwt upwards about 290.370 in number.

THE NEW WESTMINSTER BELL.

INSCRIPTION.

"This Bell weighing 13 Tons, 10 Cwt. 3 Qrs. 15 lbs. was cast by George Mears, of Whitechapel, for the Clock of the Houses of Parliament, under the direction of Edmund Beckett Denison, Q.C. in the Twenty first year of the reign of Queen Victoria, and in the Year of Our Lord, MDCCCLVIII.

NB. The Ornamentation was designed by Arthur Ashpitel, F.S.A., but in consequence of the Clock hammer being intended to strike on the outside, the lower tracery is omitted.

FIG **8.20** An artists' imagination has embellished the decoration on the bell in this print from an unknown book. It is stated that the decoration was designed by Arthur Ashpitel, FSA, but in consequence of the clock hammer being intended to strike on the outside, the lower tracery is omitted.

Since the bells were to be hauled up the central weight shaft, the order of events was fixed. First, the four quarter bells had to be raised and placed on the floor of the belfry, one at each corner. Then the great hour bell would be raised into the belfry, and finally the clock movement could be lifted into place and installed. Mr Henry Hart organized the lifting of the bells; he was assistant to Jabez James, who reported to Thomas Quarm, the Clerk of Works.

When using a steam winch to raise materials in the Victoria tower, it was noticed that the load sometimes oscillated, no doubt because of the spring in the lifting cable plus the regular pulsations of the steam engine. It was, therefore, decided that the risk of damage was such that the lifting of the bells should be done by a hand-powered winch.

Meanwhile Messrs Hawks, Crawshay, & Co. manufactured 1,500 feet of lifting chain. Each link was made of best cable bolt iron 7/8 inches in diameter; the entire chain was tested to 10 tons, and the breaking load established at 13¾ tons.

In the belfry, a substantial wrought-iron framework had been built to hang the bells. This was made of 12 cast-iron standards that stood at about 15° from the vertical, three standards on each of the four sides of the frame. The bottoms of the standards stood in cast-iron shoes that were mounted on rollers that sat on cast-iron plates set into the walls of the tower. The idea of the rollers was to allow movement of the frame brought about by vibration and changes of temperature. With such a huge weight on the tower, the walls would eventually have been forced to bulge outwards, so tie bars were connected to each shoe to prevent this happening; these went all the way around the tower and across from wall to wall. Four large wrought-iron beams were placed

in the form of a square on top of the 12 standards. India-rubber pads were placed between the tops of the standards and the beams to absorb vibration. Today all is as it was, the tie beams still tying, the rollers are there but probably have never moved, and the India-rubber pads are as hard as the iron they separate. The whole structure was secured with rivets and all the corners were braced with triangular plates. When the bells were finally raised, the hour bell was supported from two beams that bisected the outside square, and the quarter bells each hung from a smaller diagonal beam across the corners of the main square.

A massive wooden structure was built in the belfry above the bell frame. Its job was to support the winding crab that was to raise the bells. The winding crab was a winch with a large chain drum about 18 inches in diameter and 4 ft wide, this was connected to two long winding handles through reduction gears; motive power for the jack was eight men, four on each of the two winding handles. The lifting chain was wound round the barrel ten times until it gripped by friction when under load, the lifting end passed down the tower and the free end passed back into the clock room. To lift the great hour bell, the chain was passed between two pulley blocks having six falls of chain giving a total mechanical advantage of six to one. In operation, the men wound the handles and, as the barrel turned, the chain was slowly taken up and the bell lifted. As winding progressed, the turns of chain traversed across the barrel until they reached one end. At this stage the lifting chain was secured with an iron chain stopper that was specially constructed for the job. It comprised two halves that were shaped to fit the chain; these were bolted together and supported from a large wooden beam. For added security in the event

MOVING THE GREAT BELL.

FIG **8.21** The first stage of raising the bell was to put the bell in a wooden cage and move it into the base of the tower.

RAISING THE GREAT BELL TO THE CLOCK TOWER, WESTMINSTER.

FIG **8.22** Big Ben in its cage about to be hoisted up the tower.

of a mishap, the free end of the chain was taken up by another crab positioned in the clock room below. This process of moving the chain from one end of the barrel to the other was known as *fleeting*. Denison recorded that it took 15 minutes to fleet the chain across the barrel between windings.

First, the quarter bells were raised up the weight shaft. Being relatively light, the six-fall blocks were not needed.

Edmund Beckett Denison and Charles Barry failed to communicate effectively, so when it came for the bells to be specified and ordered, the tower was part finished. The only way a large bell could be raised up the tower was by the outside or up the clock weight shaft. By specifying the size of the hour bell, it was possible to use the path of the weight shaft.

On 5 October 1858, Big Ben was lowered from his wooden frame and placed onto a cradle ready

for his journey to the belfry. *The Times* referred to the bell as *Victoria*, so there was still some uncertainty as to the bell's name.

First, the massive hour bell was turned on its side and then wheeled on a carriage into the clock tower. A solid wooden cradle was then built round the bell, bolted together with iron straps, and fitted with rollers on each side. Planks had been put on the weight shaft walls for the carriage roller to run on, and a series of 12 or 14 gas jets installed to illuminate the bell's progress. At 7 ft 10½ inches in height, the bell had only 7½ inches spare when laid on its side and fitted into the weight shaft which measured 13 ft 2 inches by 8 feet 6 inches.

Lifting commenced at 6 a.m. on 13 October 1858 and continued until midday on the 14th, when the Great Bell was received into the clock room and then tilted from its sideways position to mouth down. The only breaks were at intervals of approximately 1 hour and 20 minutes for the fleeting of the chain; during each lift the bell rose some 6 feet 10 in. A rope marked with feet was attached to the bell cradle and this passed up the weight shaft and over a pulley. In this manner, the progress of the bell could be ascertained with certainty. In all, the bell was raised 175 ft to the clock room. After a weekend break, the hour bell was finally raised from the clock room into the belfry, completing its journey of 200 feet from ground level.

MODE OF RAISING THE GREAT BELL.

FIG **8.23** Shifts of eight men operating the winch handles took over 24 hours to raise the bell.

FIG **8.24** To safely raise Big Ben, a timber cage was constructed that held the bell on its side. Bell and cage together weighed 16 tons.

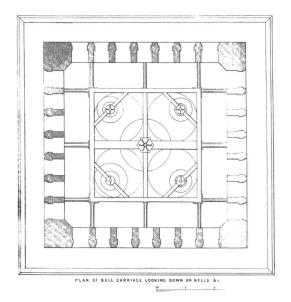

PLAN OF BELL CARRIAGE LOOKING DOWN ON BELLS &c.

FIG 8.25 A plan showing how the bells were to be installed in the belfry.

Once in the belfry, the bell was raised to within inches of the supporting frame when it was supported underneath with wooden packing. A major task then ensued; removing the lifting crab and its stout wooden support. With this completed and the belfry clear, the Great Bell was finally levered into its final resting place. On top of the Great Bell was an unusual design. The crown had a mushroom-like top. This was to support the bell when clamped by a strong cast-iron two-part collar and packed with India-rubber to give the bell freedom to vibrate. Six long bolts passed from the collar to a plate above the bell exactly on the intersection of the two cross beams. An additional bolt went through the crown of the bell to secure the clapper.

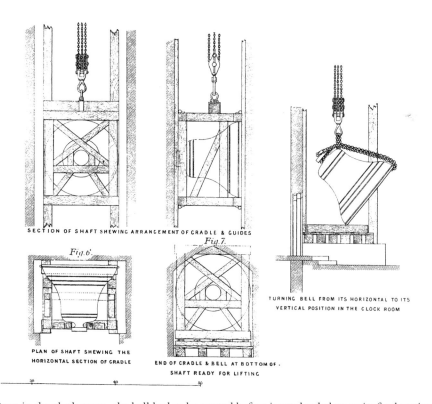

SECTION OF SHAFT SHEWING ARRANGEMENT OF CRADLE & GUIDES

Fig. 7.

Fig. 6.

TURNING BELL FROM ITS HORIZONTAL TO ITS VERTICAL POSITION IN THE CLOCK ROOM

PLAN OF SHAFT SHEWING THE HORIZONTAL SECTION OF CRADLE

END OF CRADLE & BELL AT BOTTOM OF. SHAFT READY FOR LIFTING

FIG 8.26 Once in the clock room, the bell had to be turned before it was hauled up to its final resting place.

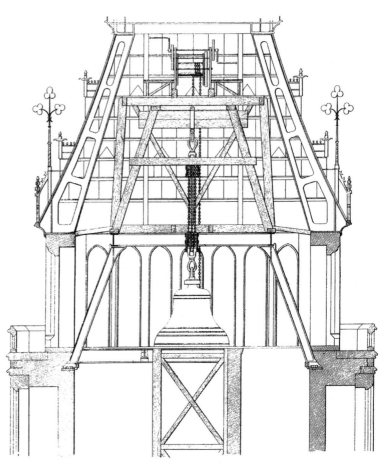

FIG 8.27 Lifting the bell into the belfry.

SECTION OF TOWER SHEWING ARRANGEMENTS FOR LIFTING BELLS.

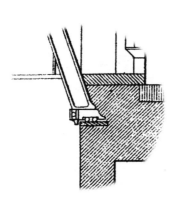

FIG 8.28 Detailed elevation of the bell frame, showing how the frame standards were mounted on rollers.

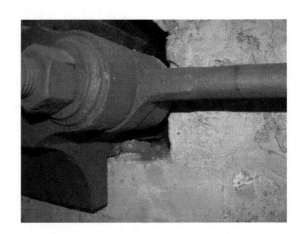

FIG 8.29 A roller below the foot of a bell frame standard.

THE NEW BIG BEN SPEAKS FOR THE FIRST TIME

At this stage, the dials were installed and the hands fitted, along with the internal motion-works. Forges still occupied the clock room, as workmen were making ironwork for the lantern. On 19 November, Messrs Denison and Taylor climbed the clock tower and a rope was attached to Big Ben's clapper and passed down to the clock room. Mr Denison set to work with a will to make the bell speak. The first stroke was slight but after that came peal after peal in a tremendous volume, which was actually painful for those present. After Mr Denison brought forth the bell's maiden speech, three workmen worked on the rope to develop the full tone of the bell. Whilst the bell was sounding, the supporting steelwork was checked and it was found that the whole structure vibrated like tightened cords and made the beams and standards work to a degree that was unpleasant to witness. It was concluded that the framework would have to be strengthened with large wrought-iron brackets. To let the world below hear the bell more clearly, it was necessary to remove tarpaulins that shrouded the openings in the belfry. A wooden roof had been constructed above the belfry, so this further altered the bell's sound at ground level.

Subsequently, the quarter bells were hung from the iron frame, hammers attached, and the bell tested. Experiments continued to establish an appropriate hammer weight and fall. Mr Quarm, unknown to Edmund Denison, had at one stage ordered the bell's rigid fixings to be slightly loosened and that improved the note significantly. Denison finally fixed on a hammer of 6 cwt 3 qrs 10 lb with a fall of 13 inches.

BELFRY IN THE CLOCK TOWER OF THE NEW HOUSES OF PARLIAMENT.

SCAFFOLDING FOR RAISING THE QUARTER-BELLS IN THE CLOCK TOWER OF THE NEW HOUSES OF PARLIAMENT.—(SEE NEXT PAGE.)

FIG **8.30** Ready to receive the bells. It was soon discovered that the belfry's slender iron framework was not strong enough to bear the weight and shock of the Great Bell sounding. From *The Illustrated London News.*

THE GREAT BELL AND THE QUARTER BELLS IN THEIR PLACES IN THE CLOCK TOWER, WESTMINSTER

FIG **8.31** With the bells hung, the frame looked decidedly flimsy.

THE CHIMES—BIG BEN AND HIS LITTLE BROTHERS

FIG **8.32** A later engraving shows that the bell frame had been fitted with massive stiffening corner braces.

FIG **8.34** The 6½ cwt hammer head that cracked the bell. This is now on display in the visitor room with the clapper ball.

FIG **8.33** Despite having no plan for operating an internal clapper, one was made and installed, but never used.

CAMBRIDGE QUARTERS
(MORE COMMONLY CALLED WESTMINSTER QUARTERS).

1st Quarter.   2nd Quarter.

3rd Quarter.

4th Quarter.

Hour.

FIG **8.35** The famous Cambridge or Westminster Quarters.

FIG **8.36** Big Ben, circa 1950; the clapper can be
seen underneath the bell.

When the hour bell was finally installed in the
tower, a report appeared in *The Manchester
Guardian* of 22 October 1858, recounting how in
a ceremony in the belfry the previous day a
workman struck the bell 21 times with a hammer.
'*Between each blow the workmen gave three hearty
cheers and Mr. Quarm the Clerk of Works designated
the bell 'St Stephen' and by that name it will hereafter
be known.*' Like *Victoria*, this name, however
appropriate, faded away as quickly as the 21
blows flew into the London air.

CHAPTER NINE
THE DIALS

EARLY DESIGNS

Barry's early design for the Houses Parliament shows a clock tower that is much smaller than the final version. Not only is it smaller but it is more ornate and is surmounted by a lantern-style bell cupola of stone tracery.

Plans in the National Archives show various developments of the dial design; unfortunately the plans are not dated. An early design shows a dial with 24 hour divisions. Such dials were used on astronomical clock dials, like those at Hampton Court, Wells cathedral, and Wimborne Minster. The dial also has an annual calendar,

phases of the moon, and a hand that was probably intended to show the age of the moon. Although it is a grand copy of a mediaeval clock dial, the excessively complex design would have been illegible from ground level. This seems to be the sort of design that Pugin would have produced.

Another later design shows something approaching the final design. Here the Roman chapters are contained inside a shield. Again, the legibility of the figures would have been poor. A section through the dial plate shows that the centre is sunken, with the hour hand travelling in the depression. Vulliamy was a champion of

THE NEW PALACE OF WESTMINSTER.

FIG **9.1** A steel plate engraving by Prior published in 1848 depicting the Houses of Parliament with an early design for the clock tower.

FIG 9.2 Detail of Prior's engraving showing a simple, uncomplicated dial.

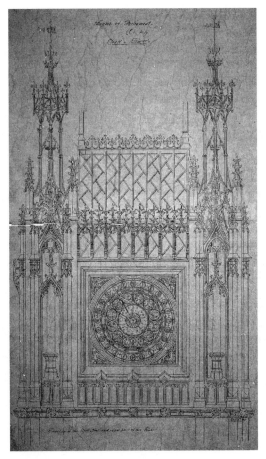

FIG 9.3 An early design of elaborate tracery for the top of the clock tower. Picture courtesy of the National Archives.

this type of dial and used it frequently. His reasoning was that this arrangement allowed the minute hand to be very close to the dial, thus eliminating the effect of parallax and allowing the time to be read more accurately. Vulliamy had worked with Barry on various projects, so a cordial communication was already established and Barry would probably have been influenced by Vulliamy's principles.

An interesting plan depicts what is closest to the final design, when the size was settled on a 22½-ft diameter. Here the draughtsman has combined four different designs for the tracery

in the centre, two shapes of minute hand, and different sizes and designs of hour chapters, as well as a selection of minute and five-minute marks.

There are also plans that show the supporting brackets that take the weight of the hands along with their attendant anti-friction roller bearings. These might date from before the time that Dent won the contract, so the design would have been influenced by Vulliamy. There seem to be no plans for the motionworks inside the clock room.

3407

FIG 9.4 Final design of the dials, laid out as a multi-choice selection plan. Picture courtesy of the National Archives.

QUESTIONS

Soon after he was appointed as referee, in January 1852, Edmund Beckett Denison started to ask questions about the dials, bells, and position of the clock. He had heard that the intention was to have metal dials and proposed that the dials should be stone and made under the direction of those responsible for the clock. As a throw-away comment, he said that the figures could be anything that Mr Barry liked, since no one looked at the figures, just the position of the clock hands.

THE TOWER GROWS

As the tower crept upwards and neared completion, the Press were ever eager to interpret what the dials would finally look like. In early 1856, an engraving of the top of the clock tower showed a dial with round chapter marks with Roman numerals. In March, questions in the House of Commons revealed that the dial plates were part of experiments with the lighting. By August 1856, another picture depicted heart or shield-shaped cartouches for the hour chapters. In 1857, a completed view of Mr Speaker's House and the Clock Tower was shown; by then the dial representation was almost correct.

CONSTRUCTION

Each dial is made of cast-iron sections; there are 12 outer sections, one for each five-minute block, and these are bolted together and bolted into the wall. Exactly how the centre part was put together is not obvious, but this too would have been made in sections, since it would be too large for a single casting and too difficult to manipulate. Once assembled, the dial would have been painted and then glazed.

THE GREAT CLOCK-DIAL OF THE NEW HOUSES OF PARLIAMENT,
ILLUMINATED.—(SEE NEXT PAGE.)

FIG **9.5** An artist's impression of the dials, from *The Illustrated London News*, 2 February 1856.

PRESENT STATE OF THE CLOCK TOWER OF THE PALACE OF PARLIAMENT, WESTMINSTER.—(SEE NEXT PAGE.)

FIG **9.6** An artist's impression of the dials, from *The Illustrated London News*, 2 August 1856.

GLAZING

An engraving of the Lighting Chamber of the Great Clock appeared in *The Illustrated London News* on 8 October 1859. The gas pipes and burners were missing but the accompanying text tells how the sketch was made before the gas lighting installation was put in. This indicates a date of late September for the gas lights being fitted. The report stated that Messrs Gardner of The Strand undertook the glazing, employing a new patented glass instead of ground glass. This glass had the appearance of porcelain when viewed on the surface but was semi-transparent when held up to the light and diffused a copious amount of light that made it suitable for its purpose. This type of glass became known as pot opal, and the milky-white colour was obtained

FIG **9.7** A detail of the Clock Tower, showing the dials, from *The Illustrated London News*, 11 April 1857.

Instead, gutta-percha was employed to bed the glass in and this was sealed with an adhesive material prepared with balsam.

As a result of a dispute over Messrs Gardners' bill, the case was referred to an arbitrator, who awarded them the sum of £3,172 17s. 9d. Considering that there were 312 pieces of glass in each dial, this meant that each pane cost on average over £2 10s. each. This sum for glazing the dials was considerably more than the clock cost and did not include the cost of the ironwork of the dials.

LIGHTING

From the beginning, gas lighting was used to illuminate the dial. A ground floor plan of the tower in the National Archives shows a gas meter room situated to the north of the weight shaft. Piping ran from the meter to the inside of the air shaft, just adjacent to where the unused original gas stopcock can be seen today. Presumably, the gas pipe ran up inside the air shaft to the clock dials. Behind each dial there was a main pipe that rose vertically; branches from this bore gas burners on swivel arms, each with its own gas tap. To access the burners, large iron staples had been let into the wall to allow a man to climb up and light and adjust the burners as necessary. Initially, the burners would have been types known as fishtail or batswing; these produced a flat flame more appropriate for illumination. Later developments in gas lighting involved the Argand lamp, which evolved from the oil lamp and incandescent mantles, which gave considerably more light and became available in the 1890s. It is not known if these were fitted to the dials.

Initially, the dials were illuminated only when Parliament was sitting; then the dials were illuminated until midnight or until Parliament

by adding an oxide. The stubbornness of the glass made it liable to fracture when being cut to the shapes needed, so upwards of 10,000 lb (almost 4½ tons) of glass were employed. Since the heat from the gas light would cause expansion of the glass, the use of normal putty was not possible as it went hard and was not resilient.

FIG **9.8** An illustration from *The Graphic* of 29 October 1887, showing the gas lighting arrangement.

FIG **9.9** Now capped off, this pipe once supplied the gas to a wall covered with burners.

rose. In 1864, there was a move to have the dials lit all night but it seems that this was not finally agreed until 1876.

To remove the fumes from the burnt gas, there would have had to have been vents above each dial, probably passing out through the floor of the gallery that runs round the belfry. Outside, the vents would have been fitted with a cowl to stop rain getting into the dial room. Today, no sign of the vents can be seen; no doubt they were blocked off when they were no longer needed. There would have to have been some way to let air into the dial rooms. The door to the dial room is the only door in the Clock Tower that is glazed and has a fan light above it. Most likely, there was once wire mesh where the glazing is now, to allow air to enter.

Cleaning the dials must have been a regular requirement, since gas fumes are corrosive and create sooty deposits. The windows that admit light into the clock room from the clock dial area are all glazed, no doubt necessary to prevent gas fumes from damaging the clock.

An interesting note was made in the Astronomer Royal's annual report on the Westminster Clock. 'The clock was affected by bricklayers work in the weight shaft in connection with electric lighting last October (1901) and that the cleaning of the clock which was thus rendered necessary had to be deferred to the Parliamentary recess.' This indicates that the plan for electric dial lighting was underway but it was to be a few more years until this was installed.

Towards the end of June 1905, a question was asked in the House of Commons about the dial illumination. Mr Hope asked the First Commissioner of Works if '*he was aware that Big Ben on the previous evening assumed a livid*

FIG **9.10** Cleaning the inside of a dial. A glass lantern slide shows two men working behind a dial. As there seem to be no electric lights fitted, the picture is pre-1905.

complexion.' Mr Crooks commented, '*It was green with envy.*' Lord Balcarres asked that notice be given of the Question. A few days later, the answer was that a new lighting system was being used. It was certainly mercury vapour lamps that were being used. These lamps give off a bluish-green light and since the eye is more sensitive to green light at low light levels, the green colour was pronounced at night-time. A report in *The Times* for July revealed an MP's suggestion that since the starboard dial was green, the port dial ought to be coloured red!

A reply to a House of Commons question in July 1905 revealed that the cost of lighting the dials of the clock electrically amounted to £242 for the year, the green light was incidental to the lamp used experimentally, and the cost of current, if it was adopted permanently, would be reduced by some 35 per cent.

After a visit to the Clock Tower, Frederick William Lavender, who was Master of University College, London, wrote an article for the *Gower* magazine. He wrote of the dial illumina-tion, '*The illuminant at present in use is Bastian's Mercury Vapour Lamp, which was described in "The Gower" for December 1906. Experiments are still being made with a view to improving the colour of the light emitted.*' Bastian lamps consisted of a sealed evacuated glass tube of about four feet in length. Inside the tube there was a steel anode and a cathode of a pool of mercury; the tubes ran at 110 volts (direct current) and consumed around 380 watts. For their time, they produced a lot of light at a reasonable cost. A deposit of iron eventually built up inside the glass tube, limiting the useful life of the lamp.

In May 1906, there was a debate concerning Palace maintenance, in which Mr Harcourt stated:

> *That the cost of lighting Big Ben was, roughly speaking, £240 a year. The green light was cheaper than the present light by about £80, but he was not prepared to revert to the experiment of last year unless he was assured that Hon. Members thought it desirable, and that the public would ultimately become accustomed to it in the interests of economy.*

This statement indicates that the mercury vapour lamps had been replaced, most probably by Nernst lamps. A Nernst lamp comprised a rod or glower of ceramic material that had been specially treated with oxides of rare elements. When cold, the ceramic did not conduct electricity but when heated it conducted to the extent that the tube would glow white hot and emit a light not unlike the spectrum of daylight. Nernst lamps were invented about 1897 and were in commercial production around 1905. They had the advantage over the then available carbon filament lamps of producing a better colour of light and having a longer life, since the ceramic did not oxidize or need a protective envelope. They were, however, short-lived as from around 1910 they were surpassed by lamps with incandescent metal filaments.

A photograph taken by the press photographer Hodsoll can be dated to about 1907. In this, we see the redundant gas tubes folded against the wall and the Nernst lamps hanging from brackets.

The lamps have a gauze-like cylinder round them; this could have been a measure to prevent moths and insects from flying into the lamp.

At some time, incandescent bulbs were installed for lighting. An article on Edmund Beckett Denison appearing in *Country Life* in 1949 showed a picture of the lighting chamber behind a dial with conventional light bulbs. An article in the 1957 *Horological Journal* states that there were ten 100-watt bulbs behind each dial

Fluorescent lights, or high-voltage cold cathode lamps, were eventually installed in 1957: a notice still on the wall of the lighting chamber warns *DANGER 1500 Volts*. This lighting was superseded in December 1994 by high-efficiency low-energy bulbs. These bulbs have no filament

FIG **9.11** Dial illumination in 1907 by Nernst lamps, from the London Metropolitan Archives.

or contacts but reply on high-frequency excitation of gas inside a bulb that has an internal phosphor coating. Apart from consuming low energy, these lamps have a long life, need no maintenance, have good colour, and their brightness is almost constant over a ten-year lifespan. The energy savings resulting from this work were initially estimated to amount to some £1,000 per annum. Replacement units were installed in 2008.

Controlling the Dial Lighting

On the Great Clock movement is a wheel that turns once in 24 hours. This wheel is fitted with a disc that has 11 slots on approximately half its periphery, but it performs no function. It is believed that it was originally planned to use the Great Clock to operate a gas switch. By having moveable sectors on the 24-hour wheel it would be possible for the wheel to operate a tap turning the gas on and off. The sectors would be altered from time to time, as necessary to give the appropriate times for lighting the dials to accommodate the seasonable variation in the length of the days of the year. Today, the dial lighting is turned on at dusk and off at dawn; it is switched manually by Palace engineering staff from the control room.

CHAPTER TEN

THE DEVELOPMENT OF THE GRAVITY ESCAPEMENT

THE GRAVITY ESCAPEMENT

The escapement is the key part of a clock that connects the train of gears to the pendulum. Working together, the pendulum and escapement define the accuracy of a clock's timekeeping. Denison's double three-legged gravity escapement, which was finally fitted to the Great Clock, produced an exceptionally accurate clock. In simple terms, Denison's escapement, along with a compensation pendulum, produced a clock that kept time to within a few seconds a week. In horological terms, Denison's escapement was one of the most significant inventions of the nineteenth century. In a subtle way, it supported the growing need for accurate public timekeeping, as society relied more often on railways for communication. From the 1860s onwards many thousands of turret clocks were made and installed in factories, churches, town halls, hospitals, prisons, workhouses, etc. Many of these clocks had a gravity escapement; the economic boom fuelled the turret clockmaking industry and the clocks installed brought order and timekeeping, and thus fuelled social and economic progress.

The common escapements used on turret clocks in the 1840s were the recoil, deadbeat, and sometimes the pinwheel, which was a version of the deadbeat. All these escapements suffered from the same problem; the effects of rain, snow, and wind on the dial meant that the driving force to the pendulum varied. As the driving force varied, so did the amplitude of the swing of the pendulum and this had a significant effect on timekeeping, causing what clockmakers call *circular error*. Another error that changes with varying pendulum arc is known as *escapement error* and comes as the combined effect of pendulum, escapement, and wheel train. The net result is poor timekeeping, and in the 1840s a turret clock would be rated as excellent if it kept time to better than half a minute a week; more probably the error was several minutes.

What Denison's gravity escapement did was to provide a constant impulse to the pendulum, irrespective of weather conditions, so that the pendulum arc remained the same; hence timekeeping was constant. This was achieved by raising a small weighted arm that was then 'lowered' onto the pendulum to give it a push or impulse. Since the arm was a constant weight and always descended the same distance under the influence of gravity, the impulse given to the pendulum was constant. In fact, two arms were used, one on each side of the pendulum.

Like most inventions, Denison's escapement evolved over a period of time. There is absolutely no doubt that Edmund was the key inventive force behind the escapement; he knew the

FIG **10.1** A recoil escapement.

FIG **10.2** A deadbeat escapement.

FIG **10.3** Mudge's gravity escapement, circa 1795.

problems that meant that turret clocks were often poor timekeepers, he knew the mathematics, and he knew the possible solution… *a gravity escapement*. However, in most matters, Edmund Denison rarely acknowledged the contributions of others; when it came to the actual making of the device there is little doubt that he had this done for him, almost certainly by the workmen of Edward John Dent and then of Frederick Dent. At this stage, it was likely that the development became a joint effort the inventor and maker working together to refine and improve the escapement, making it easy to manufacture.

Once in service on the Great Clock the escapement proved its worth and took on Denison's name. Edmund never patented his device and freely made information available through his books. It was seen by clockmakers to be so good it very quickly became the norm on all quality turret clocks made by British makers. American makers took on the escapement but it had no following on the Continent where the remontoire was in common use.

THE START

To understand how Edmund Beckett Denison produced the gravity escapement, it is necessary to look at the historical background that

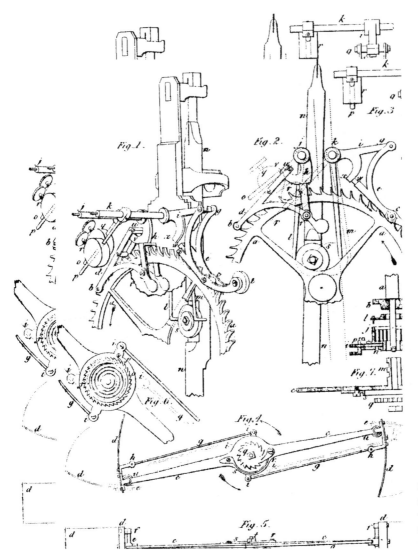

FIG 10.4 James Harrison's detached escapement, which won a silver medal and £10 from the Society of Arts in 1830.

shaped his design, and then the series of escapements that were used on the Great Clock.

Gravity escapements had been used by famous clockmakers, such as Thomas Mudge and Alexander Cumming, who had versions, and Kater, Gowland, and Bloxham invented different gravity escapements in the nineteenth century. All of these escapements were intended to go on domestic clocks, generally regulators. The escapements had the problem of being delicate and the unwary winder or user could easily get them to trip, i.e., the escapement would fail to lock and cause the escape wheel to spin round, often causing damage in the process. Whilst this was almost acceptable inside a house or observatory, such gravity escapements had no place in a turret clock inside a damp and dirty clock tower

FIG **10.5** Alkborough church clock, installed 1826.

and wound by a person who had no knowledge about clocks.

One fairly successful but little known gravity escapement was used in turret clocks, and possibly another akin to it. In 1826, James Harrison of Barrow upon Humber, bell founder and occasional clockmaker, built a turret clock for Alkborough church in the north of Lincolnshire. He was assisted by his son, also a James, and together they developed a gravity escapement, which they called a detached escapement. Colonel Quill, in his book *John Harrison: The Man who Found Longitude*, calls the father James 3 and the son James 4, in order to distinguish them from other family members of the same name. James 3 received a prize from the Society of Arts for the escapement and the description was published in the Society's *Transactions* in 1830. He was also awarded a prize for a fly with expanding blades that gave a more even speed of striking. James 3 was the grandson of James 1 Harrison, brother of the famous John Harrison who developed accurate seagoing timepieces that proved it was possible to determine longitude at sea using a clock. James 4 became a turret clockmaker, working in Hull from the late 1820s until around 1860, when he

FIG **10.6** The earliest known gravity escapement on a turret clock. Harrison's detached escapement on Alkborough church clock.

retired—he died in 1875. James 4 is known to have made at least 27 turret clocks, of which 15 had his detached (gravity) escapement.

Harrison's detached escapement had two arms, which were alternately raised by the clock mechanism and used to impulse the pendulum. These were quite complex and employed a total of eight pivots and two rollers, as compared to two pivots in a recoil or deadbeat escapement. Apart from wear, if they tripped, damage could be caused to the escapement and escape wheel. Sadly, only four of these escapements survive. Many were converted to something more conventional because of wear and because the person doing the work simply did not understand the escapement's action.

James 4 made the clock at Christchurch Hull. Its performance was exceptional and it seldom varied from its mean rate by more than second a day. This was the sort of clock that would be needed for the Great Clock at Westminster. Holy Trinity church in Hull had a clock that was improved by James Harrison in 1840. It had four dials, each of which was 13 feet in diameter and, as such, was probably the church with the largest dials in the country at that time. In his book, *History of the Town and Port of Kingston upon Hull*, J.J. Sheahan gives an extensive technical description of the clock installation. From this, it can be deduced that the four dials were driven by a special movement that was released, probably every half minute or minute, by the main clock movement. In this way, the clock was protected from the weather effects on the dials. It was, essentially, a sophisticated remontoire.

If any clockmaker in the country in the 1840s could successfully design and make the Great Clock for Westminster, it was James Harrison of Hull; but he did not live in London and was little known outside of Yorkshire.

Charles Brown of Selby made at least six turret clocks from around 1834 to 1846. The most interesting one was installed in Selby Abbey but was lost in 1902 when the tower was reduced in height because of instability, and Brown's clock was replaced by a Potts of Leeds movement. The book *Yorkshire Clockmakers*, by Brian Loomes, records an advert dated 1846 that reported:

Charles Brown, Turret Clock and Chime maker &c, Finkle Street Selby, Feels it his duty to return his grateful thanks for the continued patronage he has received during the last ten years and has much pleasure in announcing to the public that he has invented a double escapement turret clock which measures accurate time and is capable of working hands for any number of dials of any size and cannot be acted upon by irregular motion of the atmosphere. Its excellence

and durability may fearlessly challenge competition. A specimen which has been going upwards of three years may be seen in Selby Church.

This implies the date of the Selby Abbey clock to be about 1842–3. Since Selby is a little less than 40 miles from Hull, Brown must have been well aware of Harrison's work. Brown also employed features that were first used by James Harrison on his clocks. It is just possible that James 4 helped Brown develop his escapement. However, in his 1846 advert Brown confidently states that his double escapement could 'Fearlessly challenge competition,' so it is more likely that Brown based some of his clock's technical features on what James Harrison had invented. As to exactly what Brown's double escapement was remains a matter of conjecture. Sadly, Charles Brown, his wife, and two of his three children died in the Selby cholera epidemic in 1848; he was only 37 years old.

Denison's family home was at Doncaster. He was a staunchly Yorkshire man at heart, often staying at his family home in Doncaster and visiting the surrounding area. There can be no doubt that Edmund was aware of Harrison and Brown and their contribution to horology. It was quite possible that Edmund sought out James 4 and interviewed him, and probably saw the Holy Trinity clock.

George Biddell Airy was an able mathematician, and in 1826 read a paper to the Cambridge Philosophical Society on clock escapement errors. He analysed the various errors that escapements cause when driving a pendulum and concluded that the deadbeat was the best, followed by the recoil, and then the gravity. Even today, Airy's paper is still used by those keen mathematicians who want to understand clock pendulums and escapements.

Just exactly when Edmund Beckett Denison became interested in clocks is not known, but it is likely that it was long before 1838, when as a 22-year old he installed a clock at Helions Bumpstead church.

James Bloxham, also a QC, and an amateur horologist, had looked at Airy's paper and suggested to Denison that the mathematics concerning the gravity escapement was wrong. Edmund Beckett Denison presented a paper to the Cambridge Philosophical Society entitled *Clock Escapements* in November 1848. His paper was mathematical and drew on Airy's paper of 1826. Edmund concluded,

However, the object of this paper was to show that mathematically speaking, gravity escapement may be made very superior to the dead escapement with its large amount of friction and variation of arc, and to remove the cloud which has hitherto lain over them in consequence of it being supposed that whatever mechanical improvement might be made in them, they must remain liable to an insuperable mathematical objection.

A follow-up paper was read in February 1849—*On Turret Clock Remontoires*—in which Edmund describes a spring remontoire, saying that one was currently being built. This must have been the clock for Meanwood church, which was designed by Edmund, made by Dent, and installed in 1850. From these papers, we see that Edmund had a very sharp mind and understood completely the necessity in turret clocks of isolating the clock pendulum from variations in power brought about mainly by the effects of weather on the dial.

ILKLEY CHURCH CLOCK

In September 1847, Edmund Beckett Denison sent a letter to William Potts, a clockmaker in Pudsey, providing a drawing of a turret clock for the parish church at Ilkley. He wrote:

As the parishioners have desired the clock to be made according to my description, which you approve of, I thought it right to draw a more full one than the one I gave you the other day. It is the same in substance, though altered in a few particulars as you will see for the reasons I have mentioned; + I have added a few things which I daresay you will attend to without my specification, + some perhaps which you may not be acquainted with. If there is anything which you do not understand, or wish to suggest an alteration of, you must let me know.

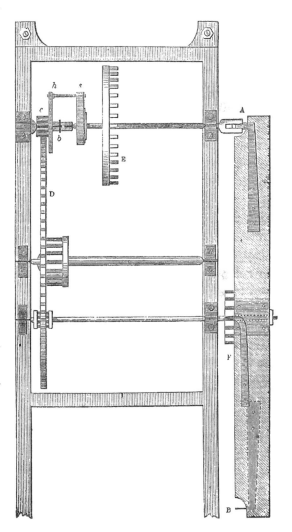

FIG 10.7 A spring remontoire, like that fitted to Meanwood clock.

The unusual features of the Ilkley clock are an expanding fly to control the speed of striking and a flirt-release striking system; these must have been what Denison referred to as things that Potts '*may not have been acquainted with.*' The significance of this clock is that the expanding fly and flirt-release had been invented by the turret clockmaker James Harrison of Hull. This indicates that Denison must have seen some Harrison clocks and would have been aware of Harrison's gravity (detached) escapement, since it was described in *The Transactions of the Society of Arts* when it was awarded a prize. Denison's knowledge of Harrison's invention must have been part of the process that brought him finally to perfect the gravity escapement as used on the Great Clock.

Denison wrote to Airy in 1846 to describe a gravity escapement. Some of his sketches appear in Mercer's book on E.J. Dent and show weighted arms that terminate in rollers, which

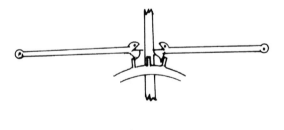

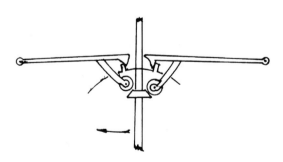

FIG **10.8** Drawings produced from Denison's sketches on the gravity escapement, made in December 1846.

roll down a slope on the pendulum rod this giving impulse. These indicate that Denison was at that time thinking along the lines of a gravity escapement.

ESCAPEMENTS FOR THE GREAT CLOCK

Detailed information can be found in Denison's book, *A Rudimentary Treatise on Clocks, Watches, and Bells*, which gives the history of the Westminster clock. Since the editions are slightly different, all eight have to be read to get a full picture. Edmund begins by introducing Charles McDowell's single-pin escapement, was patented in 1851. This invention was bought by E.J. Dent and employed in some of his watches. In the version used on clocks, a pin is fixed to the disc, which makes a half a turn with each beat of the pendulum. The advantage claimed was that the impulse was given across the line of centres; however, the disadvantage was that two extra wheels were needed in the clock train.

Dent had received the order to make the Great Clock in 1852 and within a few months the going train at least was working on a test bed. A development on McDowell's escapement was the three-legged deadbeat escapement. Denison coins the term *legged* to refer to teeth of an escape wheel and this terminology is still used when gravity escapements are described. The three-legged deadbeat escapement was the first escapement fitted to the Great Clock. It was a development of McDowell's escapement and might well have been an invention by E.J. Dent. It had a steel escape wheel that was only an inch in diameter and weighed 1/6 of an ounce. Two faces in an aperture in the crutch were used to give impulse to the pendulum and two were used for locking.

By making the locking faces with a slight curve so as to give a small amount of recoil, it was stated that the pendulum could be brought to near isochronism, i.e., having the same period for different amplitudes. The three-legged dead-beat escapement was used on the Great Clock for a period of four to six months and was super-seded by a modified version invented for the purpose of equalizing the force of impulse. This modified version might have been the improved three-legged deadbeat escapement, where some longer stopping teeth were introduced to do the locking, and the impulse teeth became shaped pins.

The similarities between the improved version of the three-legged deadbeat escapement and the single three-legged gravity escapement are striking. Virtually by replacing the pallets with two gravity arms, the deadbeat changes into a gravity escapement. It is conjecture, but one could imagine how Dent or Denison was working on the escapement with the pendulum disconnected and after seeing the escape wheel knock the pallets to one side to make the mental leap and see how the escapement could easily be modified to become a gravity escapement.

Within a year of the order for the Great Clock being placed, the going train was working on test with its third escapement, a gravity escapement.

THE THREE-LEGGED GRAVITY ESCAPEMENT

In a letter to Airy, Denison enclosed a sketch of his three-legged gravity escapement. The sketch is annotated '*Proposed gravity escapement for West-minster clock E.B. Denison, 27 Nov 1852.*' The drawing shows full-size three long teeth on the

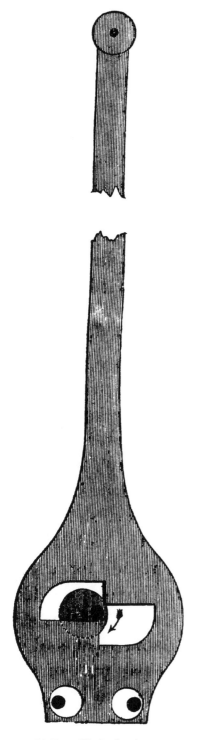

FIG 10.9 McDowell's single-pin escapement.

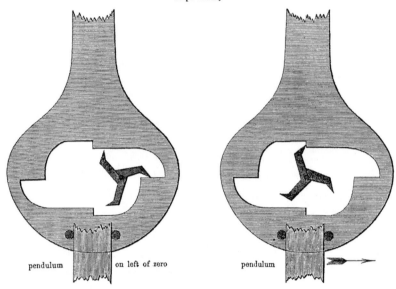

Three-legged dead Escapement: full-size for a Turret-clock; (the pallet arbor to be 12 inches above the scape-wheel.)

pendulum   on left of zero   pendulum

FIG **10.10** Denison's three-legged deadbeat escapement, as first fitted to the Great Clock.

escape wheel and two pivoted gravity arms that impulsed the pendulum. Notes give some insight into his thinking. Some formulae are provided to explain the escapement error and comments that the 686 lb pendulum requires 4 lbs falling 4 feet in a day to keep it vibrating 2°. Each gravity arm has a ball weight threaded on a stud that allows the impulse to be adjusted. A note says that the fly is behind the escape wheel and although it is not shown, the path swept by it is indicated.

In the second edition of his book dated 1855, which appeared as the section on Horology in the *Encyclopaedia Britannica*, Denison writes of the three-legged gravity escapement:

Before it was adopted for the Westminster clock, it was tried in the Royal Observatory in a common regulator, and Mr Airy, who was, as we have seen, not likely to be prejudiced in favour of gravity escapements, expressed his complete satisfaction with its performance, after trying upon it what he described as some "malicious experiments".

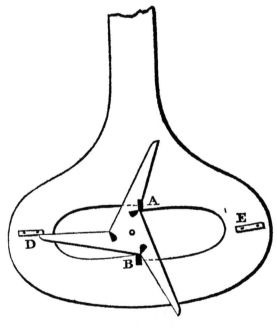

FIG **10.11** Improved deadbeat escapement. This might also have been used on the Great Clock.

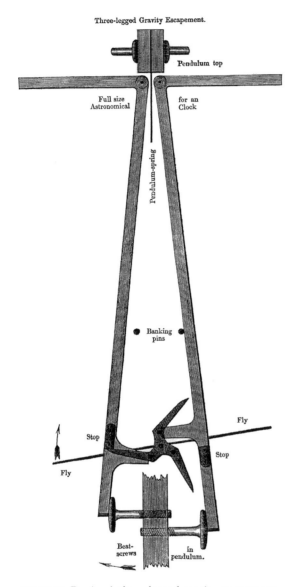

FIG **10.12** Denison's three-legged gravity escapement.

Denison's invention was very clever: by incorporating a fly on the escape wheel arbor, he solved the problem of tripping. If the escapement did trip, the fly stopped the train from running away and damaging the works. In normal operation, the fly performed another task and absorbed excess energy. The fly was mounted on the

escape wheel arbor with a friction spring. In a lightly loaded situation, the escape wheel turned smartly and when it locked the fly carried on turning for a short distance. In adverse weather the escape wheel turns in a much more sluggish manner, but when it locks, the fly again carries on but only for a tiny distance. In fair weather or foul, the gravity arms are lifted and provide a uniform impulse to the pendulum.

The disadvantage of the gravity escapement is that it needs a lot more power than other escapements. In normal operation, the unwanted power is thrown away by the fly, but when it is needed the extra power is available. Another disadvantage that was peculiar to the three-legged version was that it was prone to tripping. Under certain conditions (usually windy weather), the escapement would occasionally let two teeth pass instead of one, giving the clock a gaining rate.

Denison was consulted about a turret clock for Fredericton cathedral in Canada. The cold was so intense in the winter that oil would thicken and stop most turret clocks. The requirement to overcome this spurred Denison on to develop the gravity escapement. Regarding the practical making, he wrote, '*The best proof of the facility with which this escapement can be made, and its consequent cheapness, is the fact that two of them, one for a turret clock and one for a regulator, were made immediately from my drawings, without any mistake, by Mr Dent's ordinary workmen, in fact, chiefly by a boy; and they say it is the easiest escapement to make that there is.*' Denison also stated that several turret clocks along this plan were currently being made.

Fredericton cathedral clock was made in 1853 and must have originally been fitted with a single three-legged gravity escapement, but today it has the standard double three. Denison spoke of

FIG **10.13** The earliest surviving double three–legged gravity escapement in Fredericton Cathedral.

an excellent clockmaker in Fredericton, so it is likely that he installed the double three-legged escapement around 1859. In 1901, the cathedral suffered a bad fire in the spire, the bells were melted, and the clock damaged. However, following a restoration both clock and bells were reinstated.

A Gravity Escapement and a Remontoire

It is known that in 1856 the Great Clock had not only a three-legged gravity escapement, but also a remontoire. This was recorded by an article in the October 1856 issue of *The Engineer*. It is also known that Loseby, the chronometer maker, was complaining in 1859 that the Great Clock did not have a remontoire as the original specification demanded. The remontoire had been removed some time between 1856 and 1859. Since the great Clock was running in Dent's workshop, there was no way to see what its performance would be like in bad weather when driving four real dials. There were at least ten Dent turret clocks that had the three-legged escapement that were out in real towers telling the time, so there was a good extent of field experience of the gravity escapement. Also, Dent's prototype, described below, could have been used to test the escapement to a level sufficient to say that the remontoire was not needed.

DENT'S PROTOTYPE

What might be a prototype for the escapement is in a private collection. It was purchased by an antique dealer from Dent's when they closed down; he was told at the time that it was the prototype for Big Ben. After many years it appeared in a public auction.

This *prototype* cannot have any claim to being the prototype for the Great Clock, since it is only a single train movement and bears no common features with the movement at Westminster. However, it might well have been a test bed for the gravity escapement and has many features that argue in favour of this claim. On the other hand, there are features that cast doubt on the prototype claim.

In favour of the prototype proposal is:

- The cast iron bed is substantial and of the type used for turret clocks.

- The bed has been cut away by hand to provide the necessary openings; these were not cast in.

FIG **10.14** Was this the prototype for the gravity escapement?

- The black-blue paint looks as if it is the same as that used at Westminster.
- The whole movement has been made in a hurry with no regard for appearance.
- The wheels have not been crossed out.
- The wheels are cut with fine teeth, possibly with a cutter that was readily available.
- The great wheel appears to be a lathe change wheel, as is the first pinion.
- The pendulum is a compensation type, identical to that used by Dent before 1852.
- Arbors have been extended beyond the bearing bushes and these have cross holes
- The frame and suspension A frame have been filed to provide clearance slots for the gravity arms.
- The setting dial has minute divisions, no chapter marks, and a hole for what could be a winding hole, so it appears to be a dial reused from a domestic clock.
- The gravity arms have been fitted with outriggers onto which weights can be loaded for test purposes.

- The 1860 edition of Clocks Watches and bells reported: 'Even Mr Dent was for some time unwilling to believe that a coarse train would really answer as well as a fine one, and so he tried an experiment; and the clock with the fine train went no better, but as it happened rather worse than several of the coarse ones.' (The second and third wheels have quite fine teeth)

Against the prototype proposal is

- The workmanship is rough and not built to last.
- The gravity arms are not robust.

One interesting comment was made by Edmund Denison when he took a party of members of the British Horological Institute to visit the Westminster clock in 1875. Speaking about the gravity escapement, he said,

Old Mr Dent made a rough small clock on that principle and sent it to Greenwich for the Astronomer Royal to experiment on, and he admitted that it resisted artificial variations of force that would have disturbed any kind of dead escapement.

The most likely argument is that it is a prototype, since it has a compensation pendulum and the pivot ends have cross holes. An accurate pendulum would have been essential for any trials and experiments would have been made on loading the train; probably the reason for the extended pivots with cross holes. This is almost certainly the '*small rough movement*' described by Denison.

OTHER PROTOTYPES?

The Great Clock was running in Dent's workshop and was substantially complete by November 1853. In 1855, Frederick Dent, E.J. Dent's successor, asked to be paid for the clock. Drawings of the Great Clock are dated January 1852 and signed by the two referees, George Biddell Airy, the Astronomer Royal, and Edmund Beckett Denison. Although subsequent drawings showed alterations necessitated by the restriction on installing the clock in the tower, the clock frame and train design is essentially that of 1852.

It is not uncommon to hear a claim that a certain clock, usually in a church, is *The Prototype for Big Ben*. This usually happens when a clock is signed Dent and dated before 1859, when the Great Clock was installed. West Ham's church clock of 1857 and Cranbrook's church clock of 1856 claim to be prototypes. Since these were both built after the Great Clock was completed, there is no way that either can be a prototype. Also, Denison's drawing of November 1852 fixes a date for the three-legged gravity escapement. So these features easily pre-date Cranbrook's clock and escapement. It is not known what escapement is fitted to West Ham's clock.

Cranbrook Church Clock

St Dunstan's church at Cranbrook, in Kent, has a Dent clock of 1856, around which has grown a legend that it is the prototype for 'Big Ben'. The clock was seen by T.R.R. (Robbie) Robinson, who was an accomplished horological journalist, and who had a passion for turret clocks. He wrote extensively for *Practical Clock and Watchmaker* and when that ceased publication he wrote for the *Horological Journal*. We owe a great deal to Robbie, who wrote about so many turret clocks. Robbie's passion for turret clocks was matched by his journalistic enthusiasm. In 1967, he had a brass plate fitted to the Cranbrook clock. It does not claim that the clock is a prototype but does say that the clock had the first gravity escapement. The plate reads:

This clock embodies the actual gravity escapement invented by Mr Edmund B. Denison (afterward Lord Grimthorpe) and was used as the prototype for the mechanism of the Great Westminster Clock (Big Ben). The gravity escapement overcame the problem of supplying constant power to clock hands affected by wind and weather. Having the first gravity escapement used in a tower clock, this clock is unique.

Cranbrook's gravity escapement is, however, not the first turret clock to have a gravity escapement; that honour falls to Alkborough church, in 1826.

FIG 10.15 Denison's three-legged gravity escapement on the Cranbrook clock. Note that the place where the impulse is given to the pendulum is above the escape wheel. This has been cited as an early version of this escapement. However, the later escapement at Oakham is much the same.

The Cranbrook church vestry had a meeting on 24 November 1856, when they allocated the provisional sum of £200 for the purchase of a new church clock. Dent's tender was accepted and the total cost, including delivery and installation, was £137 17s. 3d. William Tarbutt, the author of *The Annals of Cranbrook Church*, was almost certainly a member of the clock committee. Tarbutt says in a footnote in his book that he was informed that the clock from Dent had been manufactured for the Paris Exhibition of 1855, but that it had not been completed in time and missed the exhibition. This would account for the 1856 date on the name plate.

An article about Dent supplying a new turret clock for Mexico appeared in the *Horological Journal* for October 1937. Cranbrook church clock was mentioned as having been made for the Great Exhibition of 1851; the article erroneously gives the date of the clock as 1854 and also contradicts itself over the date that the first gravity escapement was used. The historical accuracy of the article is very questionable. Considering the *Horological Journal* article and Tarbutt's comments, the most likely explanation was that the clock was made for the Paris Exhibition, the article on Dent giving only partial truths.

A man who once looked after the Cranbrook clock said that it would occasionally gain a few seconds for no apparent reason. One day he was in the clock room and actually saw the clock trip; passing two teeth instead of one. This confirms Denison's statement that this escapement was prone to tripping.

Joyce of Whitchurch's Claim to the Gravity Escapement

In his book, *Shropshire Clock and Watchmakers* (1979), Douglas Elliot records that in 1846 Archdeacon John Allen moved to the village of Prees, which is about four miles to the south of Whitchurch. Edmund seemed to have few friends and those that he did have were often clerics. Perhaps only clerics had the Christian grace to endure him. Elliot writes that Edmund was a frequent visitor to John Allen and he would drive to Whitchurch and spend a whole day in the workshops of Thomas Joyce. John Allen was the archdeacon of Lichfield cathedral, so Edmund would probably have met the man through his father-in-law, the Bishop of Lichfield.

The turret clockmakers, Joyce of Whitchurch, claim that in 1849 the company was instrumental in manufacturing a double three-legged gravity escapement after the design of Edmund Beckett Denison. The claim has been made in letters and in their website. The last of the Joyce turret clock line, Norman Joyce, died in 1966, so any family traditions and tales of the business passed down orally would still be substantially accurate. It is certain, given his interest in the subject, that Edmund Denison would have had extensive discussions with Thomas Joyce on the subject of escapements, including remontoires and gravity escapements. It is also quite possible that Thomas built an escapement model for Edmund Denison. However, nothing remains in the way of a clock, documentation, or model to give any further clue to substantiate the claim. The first single three-legged escapement can be accurately dated to 1852 and the double three-legged escapement did not appear until 1859.

Joyce's claim that they made a double three-legged escapement in 1849 to Denison's design presents an interesting point. There is a Joyce clock in Grosvenor Street in Chester, which used to be an old bank building; this has a double three-legged gravity escapement and was installed in 1853. A stately home just outside Whitchurch has a small single three-leg gravity escapement dated 1856. It is easy to update a single three-legged escapement to a double three by changing the escape wheel and gravity arms and this can be done without leaving any trace of what was there before. So a possible explanation is that Denison influenced Joyce to make some clocks to his design; most likely the Grosvenor Street clock was altered at a later date from a single three-legged to a double three-legged escapement.

Denison had a strong character and how he related to clockmakers is difficult to guess. He had got on well with his village carpenter tutor and he had a good working relationship with Old Mr Dent and his managers. Quite probably, he had sufficient character both physically and mentally to assume that anyone would help him. His social status meant immediately that he was in a superior position to the craftsmen who were probably made to feel obliged to help him

Dent produced a number of turret clocks between 1853 and 1859. These all follow the same construction and some have three-legged gravity escapements. Examples are known at:

1853	Fredericton cathedral, New Brunswick,
1853	Lode, Cambridgeshire,
1853	Northchurch, Hertfordshire,
1854	Milton Bryan, Bedfordshire (undated— date from parish records),
1856	Cranbrook,
1857	Balmoral Castle,
1857	West Ham,
1857	Enstone, Oxfordshire, F. Dent, successor to E.J. Dent,
1858	Doncaster,
1858	Oakham,
1859	Leeds Town Hall,
1860	Arlesey, Three Counties Asylum (Fairfield Hospital) Bedfordshire, Dent (date assumed from the date of the bell,
Undated	West Quantoxhead.

FIG 10.16 At Doncaster, the three-legged escapement gravity arms impulse the pendulum below the escape wheel.

FIG 10.17 Oakham's escapement is very much like that in Cranbrook, but dates from two years later. This rather nullifies the argument that the Cranbrook escapement is an early version.

A Gravity Escapement and a Remontoire

The article in *The Engineer* for October 1856, explains that a remontoire was installed so that the external hands had a discernable motion. In the 1857 edition of *A Rudimentary Treatise on Clocks, Watches, and Bells*, Denison wrote,

It was also intended to retain the visible motion of the minute hand at every 30 seconds… But when the momentum of these ponderous hands, jumping 7 inches on the circumference of the 22½ feet dial was tried, it was not considered safe to retain that motion; and the gravity escapement having been invented by that time, and in use for several years, and being sufficient to prevent the variations in force and friction in the train from reaching the pendulum and disturbing the time, the train remontoire was removed.

These comments indicate that a test was made with the Great Clock driving the hands, probably one set, whilst the clock was being tested in Dent's workshop.

THE FOUR-LEGGED ESCAPEMENT

Since the three-legged escapement was known to cause problems by tripping, a four-legged one was developed. It seems that, this was not used at Westminster but Dent did employ it this in domestic regulators, the accurate clocks beloved of astronomers and clockmakers. Denison had a regulator made for him with a four-legged escapement. In 1902, he wrote fondly of his clock.

I am sitting now in front of the very first of them, which was made straight off from my drawing in 1852, and has been going ever since, either in London, while I kept a house there, or here. I am sorry to say that the actual maker, James Brock, is dead. He was an excellent and charming man, who first worked for me at the original Dent's, in the Strand, and built a sufficient factory in a stable-yard in that region, where we made the Westminster clock and sundry other large ones.

The four-legged version was used on domestic regulators that had a gravity escapement, since for a one-second pendulum; the gearing to drive a seconds dial is more easily achieved than with the double-three type. Turret clocks with the four-legged gravity escapement are uncommon

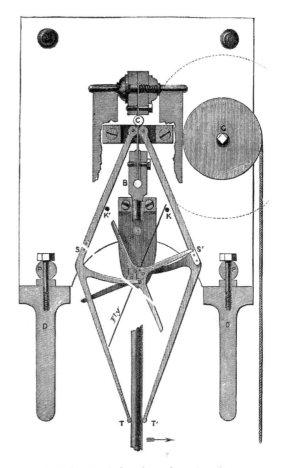

FIG 10.18 Denison's four-legged gravity escapement, from *A Rudimentary Treatise on Clocks, Watches, and Bells.*

FIG 10.19 A four-legged gravity escapement on a flatbed turret clock movement by Joyce of Whitchurch of 1884.

but Joyce of Whitchurch made some movements on this plan from the late 1870s to the 1890s.

THE DUAL THREE-LEGGED GRAVITY ESCAPEMENT

Like most developments, Denison's escapement evolved in stages. At one stage of its life the single three-legged escapement on the Westminster clock had two escape wheels, probably making it a dual three-legged gravity escapement. No examples of this escapement are known and only a fleeting reference in the 1857 edition of *A Rudimentary Treatise on Clocks, Watches, and Bells* gives information as to its existence. Denison wrote,

And in order that the action may be in the direct line of both pallets, notwithstanding their making a greater angle with each other than usual, on the account of the length of the teeth, there is a double escape wheel; that is to say, two three-legged wheels set on the same arbor about half an inch apart, with one set of pins between them to lift the pallets; and one wheel locks on a stop in front of one pallet and the other on a stop on the back of the other pallet.

It is difficult to follow exactly what Denison was trying to achieve, but his description of two separate escape wheels is perfectly understandable.

It would seem likely that this escapement was the one fitted to the Westminster clock when it was installed in 1859. Once Denison's ideas on locking to eliminate tripping were formulated, it was a short step to modify the dual three-legged escapement to make it a double three-legged one.

DENISON'S DOUBLE THREE-LEGGED GRAVITY ESCAPEMENT

The final and successful escapement that was fitted to the Great Clock in 1860 was the double three-legged gravity escapement. Tripping was a known problem with Denison's first gravity escapement; his solution was to change the geometry, the size of the gravity arms, and the locking.

First, the locking blocks were moved much higher up the gravity arms to positions 60° each side of the vertical rather than on each side of the escape wheel, as was used in the three-legged version. The moving of the locking blocks' positions presented a problem: when one tooth of the escape wheel was locked on the left-hand gravity arm, the locking block on the right-hand arm would foul on an escape wheel tooth. The solution was simple: provide two separate escape wheels, one for each gravity arm. Next, the locking blocks were placed on the front of the left-hand gravity arm and on the rear of the left-hand gravity arm. The references to left- and right-hand refer to the escapement as normally viewed on a turret clock with the escape wheel rotating anticlockwise.

Second, the gravity arms were made much longer than had been generally used on the single three-legged escapement, so they had more inertia for the same weight. Having more inertia, it was much more difficult for the arm to be knocked high enough so that the clock unlocked and the escapement tripped. This was doubly the case, since the locking blocks were now closer to the gravity arm pivot points. Denison commented that long gravity arms are better than short ones; attributing the reason to clearance in the pivot holes having a much lesser effect on a long gravity arm than on a short one.

One can see how the mechanically minded could make the leap from the three-legged deadbeat to the three-legged gravity escapement. But the development of the three-legged gravity escapement to form the double three-legged version could only be the result of logic and scientific thought. Denison is the only person who could be credited with such facilities. It is known that the double three escapement was fitted first either to Leeds town hall clock or retrofitted to the clock in Fredericton cathedral. Since development and experiment work would have been necessary to test the invention, this further strengthens the case for the prototype movement.

This final version of Denison's escapement was developed around 1858–9 and became known as Denison double three-legged gravity escapement. The double three-legged refers to the escape wheel being two three-legged wheels.

Exact dates do not exist, but we know that by mid-1859 the Leeds Town Hall clock was started, and this was fitted from the start with the new double three-legged escapement.

Once the Great Clock had been running, Airy made an inspection of the clock and bells and wrote a report. Although he had managed to sever his duties as a referee because of working with Denison, Airy was retained as a referee by the Commissioner to report on the Great Clock and bells.

Airy's report was dated 21 April 1860, and he wrote:

I understand from him [Denison] that there is now a fear of the clock tripping under the effect of sudden gusts of wind acting on the hands. (The actual escapement is liable to the defect of 'tripping', that is, to the train of wheels running forward without corresponding vibrations of the pendulum, upon receiving a strong impulse.) Mr Denison has prepared another escapement (which I have not seen) intended to be free from liability to this inconvenience, and which, as I understand, is ready for mounting, and can be mounted in one day. I recommend that authority be immediately given to Mr Denison to mount the new escapement.

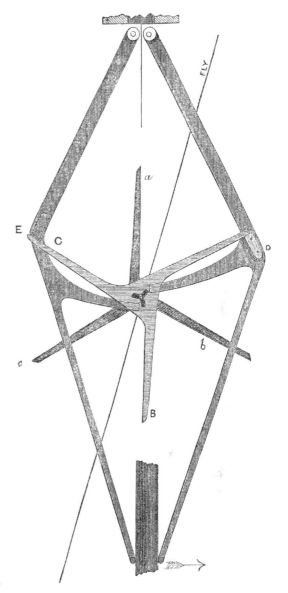

FIG 10.20 Denison's double three-legged gravity escapement, from *A Rudimentary Treatise on Clocks, Watches, and Bells.*

There is little doubt that the invention of the gravity escapement was not a flash-in-the-pan idea but a long, hard process, starting with an objective, and followed by a series of ideas, experiments, testing, modifications, and development.

FIG **10.21** The escapement from the Great Clock, in the workshops of the British Horological Institute at Upton Hall.

THE DEATH OF FREDERICK DENT

On 25 April 1860, Frederick William Dent died; he was just 52 years old. Frederick had been suffering with *delirium tremens* (DTs) before he died; his death certificate recorded 'congestion of the liver'. The cause of death combined with the DTs implies that he had been a heavy drinker for some years.

On 10 March, Frederick signed a will that Edmund Denison had drawn up for him; in this Frederick chose Edmund Denison as his executor and left the turret-clock business to him. The will was missing after Frederick's death and it was claimed by the family that Frederick ordered the will to be destroyed shortly before he died. Denison disputed this and a court case between Denison and Elizabeth Dent, Frederick's mother, was heard in December 1860. The court found in favour of Elizabeth Dent, and Denison received nothing. Legally, this meant that Frederick had died intestate and the business was passed on to Elizabeth Dent, its new

name becoming first Dent & Co., and then E. Dent & Co.

It was a difficult situation: a will had been made by Frederick, who was rational one minute and not the next; his mother, Elizabeth, no doubt wanted to preserve the business that her husband and son had built up; Edmund the lawyer always did things to the absolute letter of the law, but failed to see that he was personally involved and too close to the family. The net result of the dispute was that Denison did not work with Dent after Frederick's death. The company continued to maintain the Great Clock. Having lost Frederick, the mechanic, and Edmund, their chief mentor, it seems that Dent's output of turret clocks slowed down before picking up in the 1870s. Denison was never slow in criticizing someone he thought was wrong; however, he never said anything against Frederick Dent and, indeed, Frederick Dent's name appears as the author on a rare version of Denison's *A Rudimentary Treatise on Clocks, Watches, and Bells*. The second edition formed the text on horology for *Encyclopaedia Britannica* and one printing of 1855 was entitled *Treatise on Clock and Watch Work* by Frederick Dent.

LEEDS TOWN HALL

Leeds Town Hall ordered a turret clock from Dent in May 1858; Warners were to make a bell of 35 cwt. By February 1859, a heavier bell had been specified and Dent was to charge £550 for a clock to suit the new bell. This change may possibly have delayed Dent from starting work, or from modifying the clock if work had started.

In June 1859, Alderman Kitson of Leeds visited Dent's workshop and found that little progress had been made on the Town Hall

FOREWORD PLATE Mike McCann, MSc, BEng (Hons), CEng, MInstE. The Keeper of the Great Clock.

PLATE I **FIG 1.3** The great bell 'Big Ben'.

PLATE 2 **FIG 2.16** Dial with inscription, probably devised by Pugin.

PLATE 3 **FIG 2.18** The Great Clock, showing the flies above and the connections to the dials.

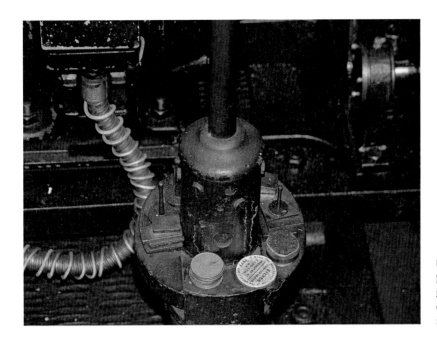

PLATE 4 **FIG 2.27** Add one penny and the clock will gain 2/5 of a second per day; taking a coin off makes the clock lose.

PLATE 5 **FIG 2.28** Denison's double three-legged gravity escapement.

PLATE 6 **FIG 2.29** Winding the going train by hand.

PLATE 7 **FIG 3.3** The inside of Westminster Hall, showing the magnificent hammer-beam roof. Etching and aquatint by Rowlandson and Pugin produced for the 'Microcosm of London' series, originally published 1808–10.

VIEW of the HOUSES of LORDS & COMMONS, destroyed by FIRE on the night of the 16ᵗʰ October 1854.

PLATE 8   FIG 4.3 A chromolithograph, captioned 'View of the Houses of Lords & Commons, destroyed by Fire on the night of 16 October 1834'. Sketched from Serle's Boat Yard.

PLATE 9   FIG 6.7 A wax portrait of Benjamin Lewis Vulliamy, Clockmaker to the Queen.

PLATE 10   FIG **8.18** A wooden profile of Big Ben inside the Foundry's door. The author inside is 6 feet tall. With thanks to the Whitechapel Bell Foundry.

PLATE 11   FIG 12.1 The two cracks in the great bell: the closer one in the bright sunlight is the longer, the further one is the deep crack.

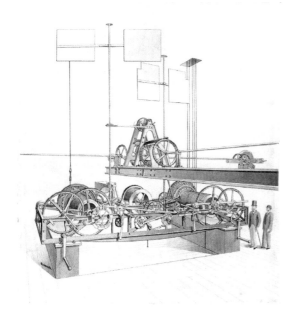

PLATE **12** FIG **13.5** Dent's drawing of the Great Clock.

PLATE **13** FIG **13.6** *Bells:* The *Vanity Fair* cartoon of Lord Grimthorpe by 'Spy' that appeared on 2 February 1889.

PLATE 14 **FIG 13.7** A pastel portrait of Lord
Grimthorpe.

PLATE 15 **FIG 19.3** Left to
right: in the background,
Huw, Paul, and Ian,
Clockmakers at the Palace of
Westminster. In the front,
three visitors to the show:
Callum and Lewis are
holding the bronze spanner
used to remove the octagonal
retaining nut on the minute
hand, while Tyler holds the
winding handle employed to
wind the going train.

clock, so an ultimatum was issued that unless significant advance was seen within a month, the order would be cancelled. Quite possibly, the lack of activity was a result of Dent's deep involvement with the installation of the Great Clock from April onwards. It is also possible that Dent's health was at this time severely affecting his performance. If this was the case, then it would imply that Denison was the person who was managing the Great Clock's installation.

By October, the building of Leeds Town Hall had progressed sufficiently to be able to receive Dent's clock, but the clock was not installed until 1860. William Potts was an up-and-coming clockmaker in Pudsey; he made turret clocks. Potts, who was soon to establish himself in Leeds, recorded in his note book that work had commenced on the Town Hall clock on 8 June 1860, so the installation was probably completed just before this. Potts was tasked with the winding and full care of the clock from 30 June. Since Frederick Dent had died in April, the Leeds clock would have been installed by Dent employees. Denison would not have been involved, since he was disputing Frederick's will and no doubt would have been unwelcome in the Dent organization.

In September 1860, Denison wrote to Potts outlining the problems with the clock. He was defensive of Dent and outlined the unsatisfactory way in which the architect failed to accommodate proper supports for the leading-off and motionwork. Potts spent a long time getting Frederick Dent's last clock right.

Denison had worked with William Potts of Leeds in the past, on the Ilkley clock, and must have been drawn into a closer relationship with him, aided by problems with the new clock in the Town Hall. Potts started to use Denison's double three-legged gravity escapement; the first clock he made with it was for Bradford cathedral in 1862. Connections with Potts were to last a lifetime; in his will Denison left a clock to Trinity College, Cambridge. Potts was one of the three companies that were specified as a suitable maker.

Today, Leeds Town Hall clock is still the flatbed clock supplied by Dent, dated 1859. The clock is distinctly undersized to drive four dials, each of which is 10 ft 6 inches in diameter, and to strike the hour on a bell of 81 cwt. The only difference between the Leeds clock and similar clocks by Dent is that the striking great wheel is larger than normal and part of the frame has been cut away to enable it to fit.

Certainly, Leeds had an early double three-legged escapement but this is now lost, since Potts replaced the escapement with their double three-legged gravity escapement in 1881. Potts frequently replaced worn or unfamiliar escapements on other turret clocks; presumably he found this more reliable then effecting repairs. As the Leeds clock is rather small for the dials and there were numerous problems with the bevel gears and leading-off work, most probably the clock had extra driving weights added, which would have accelerated the wear on the escapement. Also, since the Dent escapement was an early version of the double three-legged escapement, it might not have been as robust as later versions. Denison specified that the escape wheel of a gravity escapement should not be fixed with a pin but put onto a spline on the escape wheel arbor. Perhaps this observation came from experience on the Leeds and other similar clocks.

At some time, the hands were altered by riveting Potts-style hands onto the original Dent

FIG 10.22 Leeds Town Hall clock was made by Dent in 1859 and installed in 1860.

FIG 10.23 Replacement escapement by William Potts, installed in 1880.

hands. A night-silencing device was added, possibly at the same time that the escapement was altered; this meant that the clock was silent after striking 9 p.m. and until 7 a.m. the next morning, when striking recommenced. Potts also installed automatic winding on the clock, probably in the 1970s, discarding both the striking and going barrels.

DENISON REMEMBERS THE FIRST DOUBLE THREE-LEGGED GRAVITY ESCAPEMENT

In his letter to the *Horological Journal* of 1902, Denison wrote:

> *Sundry others of both kinds, four-legged and double-three, were made, of which I have no list nor adequate memory. But I can name Manchester and Bradford and Rochdale and several other town halls; and Fredericton cathedral, in Canada, which was finished first of all; or, perhaps it was Leeds town hall clock with the double three legs.*

In saying that one of the first double three-legged gravity escapements was fitted to the clock in Fredericton cathedral, Denison implies that the single three-legged escapement originally fitted was replaced. We know that there was a competent clockmaker, who looked after the cathedral clock, and who probably did the job under instructions from Dent and Denison. In the 1860 edition of *A Rudimentary Treatise on Clocks, Watches, and Bells*, Denison wrote,

> *Fortunately there is a scientific clockmaker there called White, who has taken great interest in it, and shown considerable judgement in making various little alterations which were naturally required in a machine of new construction, not even fixed by the maker; and I have heard from him very satisfactory reports of its performance, even when the oil has frozen as hard as tallow.*

LATER TYPES OF GRAVITY ESCAPEMENT

Clockmakers were always trying to improve escapements. Variants of Denison's gravity escapement included one with five legs that Cooke of York used in conjunction with a remontoire, much to Denison's scorn. Thwaites and Reed have a six-legged version and Gillett & Co. made several types, including a 15-legged escapement and a six-legged one that had one impulse arm and two locking arms. Joyce of Whitchurch made a number of clocks with the four-legged gravity escapement. One bizarre invention was by W.G. Schoof, who proposed a turret clock having his gravity escapement, an escape wheel with 3,600 teeth and a pendulum 120 feet long. Clockmakers and amateurs all produced different types and these were frequently reported in the *Mechanics Magazine* and *Horological Journal*.

Gravity escapements were invented by:

Clarke,
Cooke,
Cunninghame,
Dr Waldo,
Gillett and Johnston,
Granger,
Higham,
Hutchinson,
Kater,
Kempe,
Lange,
Leeson,
Ley,
Peers,
Pyke,
Schoof,
Thwaites and Reed,

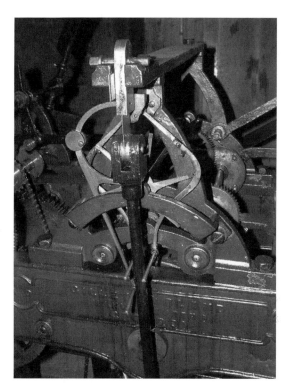

FIG **10.24** A six-legged version of the gravity escapement by Gillett and Johnston.

Vivier,
Younghusband.

THE DOUBLE THREE-LEGGED GRAVITY ESCAPEMENT IN OPERATION

The objective of the gravity escapement is to ensure that variations in force (caused by wind and weather) that reach the escapement do not reach the pendulum.

The principle of a gravity escapement is that two gravity arms are alternately raised to a fixed height and, when unlocked, descend a fixed distance and impulse the pendulum.

Since the impulse is always produced by an arm of a fixed weight descending a fixed distance, it remains constant. Since the impulse is constant, the pendulum arc is constant and circular errors are not produced, so time-keeping is good.

The escapement comprises:

- Two gravity arms,
- Two escape wheels, each having three legs (teeth),
- Three lifting pins, which separate the two escape wheels,
- Two locking blocks, one on each gravity arm,
- A fly that can rotate on the escape wheel arbor but is restrained by a friction spring,
- A pendulum.

In operation, the gravity arms operate as illustrated in the diagrams. In Figures 10.26 and 10.27, the escape wheel turns anticlockwise and the fly has been omitted for clarity.

Figure 10.28 shows the stages of the pendulum being impulsed. Only the left- hand gravity arm is shown, the same series of events is carried out by the right-hand arm.

A. The arm is in its reset state. The pendulum approaches from right to left.

B. The arm is contacted by the pendulum. The arm is lifted and unlocks the escape wheel.

FIG **10.25** The double three-legged gravity escapement: A, gravity arms; B, escape wheels; C, lifting pins; D, locking block; E, fly; F, pendulum.

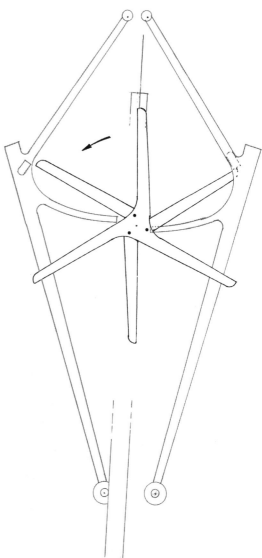

FIG **10.26** The pendulum is at the extreme left. The rear escape wheel locks on the right-hand gravity arm.

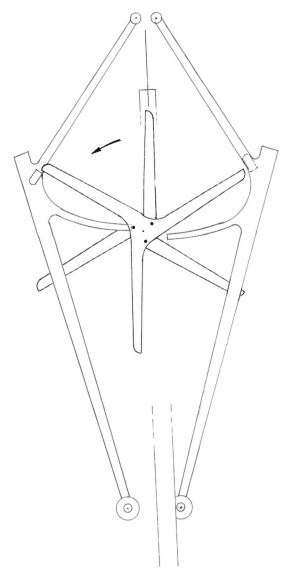

FIG **10.27** The pendulum is at the extreme right. The front escape wheel locks on the left-hand gravity arm.

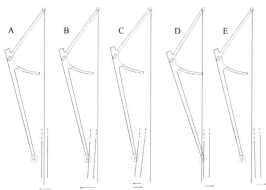

FIG **10.28** The stages of the pendulum being impulsed.

E. The pendulum contacts the right-hand gravity arm and unlocks the escape wheel, which then lifts the left-hand gravity arm to its reset position.

As the energy to the escape wheel varies, the energy beyond what is needed to turn the hands and lift the gravity arms must be lost somewhere. The energy loss is accomplished by a fly on the escape wheel. Actually, the fly does not work as an air brake but rather as a flywheel. When energy is excessive, the escape wheel accelerates quickly and locks; the fly then carries on turning and the energy is lost in the fly's friction spring. When energy is being taken by the dial, because of strong winds affecting the hands, the escape wheel accelerates slowly and the fly turns more sluggishly and does not run on so far. In normal operation, the fly will run on by 1/6 to 1/12 of a turn for every tick.

William George Schoof was an eminent chronometer maker working in Clerkenwell in London. His book *Improvements in Clocks and Chronometers* was published in 1898. In this work, Schoof wrote:

Referring to the double three-legged gravity escapement so much in use in large church clocks, its working may be likened

C. The pendulum swings to the extreme left of its swing, carrying the gravity arm with it. It then reverses direction.

D. The pendulum passes its midpoint. The falling weight of the gravity arm impulses the pendulum.

to the doings of a captain taking 30 tons of coal on board, for a trip that only requires one ton, and engaging a man called "fly" to throw the other 29 tons overboard. Substitute units of energy for tons of coal and the analogy is complete.

ESCAPEMENTS ON THE GREAT CLOCK

The Great Clock has had a series of escapements fitted to it that reflect a course of development. It is not clear exactly what was fitted and when, but the table below gives a fair representation of the progression.

THE COMPENSATION PENDULUM

Just as the escapement was important for good timekeeping, so was the pendulum. Most pendulums on turret clocks before 1840 had either a rod of iron or one of wood; the effects of temperature were simply ignored. This shows how traditional and resistant to change turret clockmakers were. Compensation pendulums to correct for the effects of temperature were used on precision regulator clocks, but not in a clock tower despite the fact that variations of temperature were far

Date	Escapement	Remontoire
1852	Denison's three-legged deadbeat	No
1852	Probably for a short time Denison's improved three-legged deadbeat	No
Late 1852–3	Three-legged gravity	No
Pre 1856	Three-legged gravity	Yes
Pre 1857	Three-legged gravity with two escape wheels	Yes
1857	Three-legged gravity with two escape wheels (dual three-legged)	No
1859	Great Clock installed in tower, probably with three-legged escapement	No
1860	Double three-legged escapement installed	No
1877	New escape wheel fitted	No
1976	Spare escape wheel fitted	No
2007	Escapement restored	No

wider than in a domestic environment. Dent was the first maker to improve timekeeping by routinely fitting compensation pendulums to his turret clocks. This move reflected Dent's nature of wanting to solve problems: an attribute that was a key factor in winning the Great Clock contract.

The principle of a compensation pendulum was to use two different metals that had different rates of expansion. One rod of metal would expand down, and the other rod that supported the pendulum bob expanded upwards. If the lengths and expansion rates were chosen correctly, the net result was that the pendulum bob stayed in the same position, irrespective of temperature. Such an arrangement was impractical with a length of metal hanging down below the bob, so makers hit on the idea of using several stages to keep a compact design. John Harrison was the inventor of the compensation pendulum; his implementation, which uses brass and iron rods is often called a gridiron pendulum. Early Dent turret-clock pendulums comprised an iron rod that expanded downwards; on this sat a zinc tube that expanded upwards. On the top of the tube was a crosspiece and two iron rods that went downwards, from which the pendulum bob was suspended. By using zinc, which has a large coefficient of expansion, the pendulum had fewer stages than a brass and iron gridiron.

This construction was not too convenient and in 1845 Vulliamy delivered a paper to the Institution of Civil Engineers on the subject of the construction and regulation of clocks for use on railways. Dent was present, and in the discussions that followed Dent said, '*He was induced to believe that a very good pendulum could be constructed, by a combination of a zinc tube ½ an inch in diameter, with a small steel rod. The expense would not be considerable, and he thought the effect would be good.*'

Also, speaking about a conventional pendulum that had been tested on a clock, he remarked, '*By careful analysis, it had been shown that of a loss of 12 seconds in a given time, 8½ seconds were attributable to the elongation of the rod, and the remaining 1½ seconds to the decreased elasticity of the spring.*' Again, this shows how Dent was operating in a scientific manner beyond that of the conventional clockmaker.

Dent's comments in February 1845 indicate that his proposed pendulum of a tube of zinc with a rod of steel was then only an idea; it was a few years before it was transformed into a reality. Dent installed clocks in churches at Tiverton in 1849 and Meanwood in 1850. Both employed a compensation pendulum of concentric tubes. Some Dent clocks made after 1850 also

FIG 10.29 The wooden pattern for the massive pendulum bob of the Great Clock.

used his gridiron zinc and steel compensation pendulums, perhaps indicating that the concentric tube version was still being evaluated.

The pendulum for the Great Clock was one of the earliest of a new design using concentric tubes. A central rod of iron descended from the suspension spring and a tube of zinc sat on a collar at the bottom. An iron tube descended from the top of the zinc tube, and the bob was suspended at the bottom of that tube. Holes in the tube allowed access for the air to enable the

FIG 10.31 Dent's compensation pendulum as fitted to the Great Clock. Photo by Hodsall, from the London Metropolitan Archives.

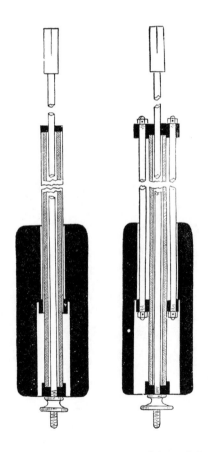

FIG 10.30 Two compensation pendulums: left, the concentric tube type as fitted to the Westminster clock; right, the type Dent used on his early turret clocks. Engraving from *Watch and Clockmakers' Handbook, Dictionary, and Guide*, by F.J. Britten (1911).

compensation pendulum to react to rapid changes in temperature. The top collar on the Westminster pendulum formed a convenient platform to hold regulation weights.

The pendulum for the Great Clock was first calibrated to time by fitting it to a common church clock in Dent's workshop that had a pinwheel escapement. Once brought to time, it was moved to the Great Clock.

In the 11th edition (1915) of his book *Watch and Clockmakers' Handbook Dictionary and Guide*,

F.J. Britten reports that so many weights had been added to the Westminster pendulum that these were removed and a new weight cast and added. Britten suggests that the zinc tube was compressing under the action of the weight of the bob. There is a large weight on the pendulum collar but this is more likely to be the weight that, if removed for 15 minutes, causes the clock to lose one second. In 1944, the suspension spring broke, so whatever weights were on the pendulum then would have been changed.

CHAPTER ELEVEN

THE GREAT CLOCK IS INSTALLED AND SET GOING

INSTALLATION OF THE CLOCK MOVEMENT

Once the bells had been raised up to the belfry and bolted to their supporting frame, other work had to be completed before the clock could be installed. The floor of the belfry had to be laid and the floor of the lever room and ceiling of the clock room built. Only when this building work had been completed could the parts of the clock be hauled up the weight shaft. Planking had been used to line some of the weight shaft to facilitate the passage of Big Ben; this all had to be stripped out.

Mr Denison had occasion to answer a question from the House of Commons regarding the progress of the clock. His answer was given on 3 March 1859. Denison wrote:

The answer to the first part of the question you have sent me is that the bells have been fixed in the Clock Tower; but they are now let down a little, to enable some pieces to be put into the frame, of the nature of diagonal braces to strengthen it against the shake caused by the blow of the great hammer. The clock is not fixed because the clock room will not be ready for it until the weight shaft is properly covered over, and the pendulum-room constructed in it, and an enclosed passage made through the air-shaft to give access from the clock room to the west dial works. This must also be made of iron, because the air-shaft is, in fact, a chimney: and it must be made carefully, so as to be air-tight, or else the fumes of the ventilating fire will come into the clock room. I am
assured that the clock room will be ready for the clock in a week, and that the bell frame will be finished very soon, and therefore I see no reason why the clock should not drive the hands and be striking in a few months.

The double-barrelled crab was erected on the two main iron beams, running east–west, and directly above the weight shaft; using this, the parts would have been hoisted up to the clock room one at a time. Probably, the first to be lifted were the two massive frame members, each of which must weigh about a ton. With the main clock frame in place the wheels and barrels could be installed without further ado.

By April, things were progressing and the clock movement was being installed, as recorded by a negative article in *The Times*. They pointed out that a means to prevent the hammer head resting on the Great Bell had not been devised and that the winding of the clock would require 11,500 revolutions on the winding handles every three days, for which they estimated that in a year, three months would be spent winding the clock. For this reason, hydraulic winding was being considered.

There were reports that inferred that the delay in installing the clock was Dent's fault. In a letter to *The Times* dated 25 April 1859, Frederick Dent wrote in response that despite the hands having been on the clock for two years, he was only notified in mid-March that the tower was ready for the clock. Within two days of being informed that the tower was ready, parts of the clock were taken up the tower; however, when he and Denison inspected the clock room they found that the place was so dark it was impossible to do any work, so windows had had to be pierced in the tower wall to let in natural light from the dials; this work took about a month. The work of the bricklayers and plasterers was finally completed in the week commencing 18 April; only then was the clock room ready to receive its occupant.

Dent also pointed out that the clock could have been installed two years earlier had the bells been ready. He went on to blame the architect for the delay in strengthening the bell frame and producing the device to catch the hammer, which had been ordered four months previously. Regarding winding, he said that hand-winding was the safest and cheapest.

Raising the parts and the assembly of the movement would probably have taken about two weeks, with another week to get the clock keeping reasonable time. Such a timescale would have meant that the clock would have been working inside the tower during the week commencing 9 May and probably brought to correct timekeeping a week later, by 16 May.

THE CLOCK IS SET GOING

Once the Great Clock was installed in the tower, it would have been set up in stages. First, the clock would have been run with no dials or bells connected. Then, once the clock was keeping time, one dial would have been connected and run for a few days. Another dial would then be connected, and so on. In this controlled manner, each dial would have been tested. It is probable that few people would have noticed a single dial

running. We do not know if, at the time that the dials were set going, scaffolding, and tarpaulins still surrounded the dials, as they did surround the belfry. A staged plan for installing the whole clock would eventually have the clock driving all the dials. It must have been in this period that it was discovered that it was only possible to drive two dials. It turned out that the hands were too heavy but this did not seem to be a real problem, since the hands would have been carefully counterbalanced. The real problem was the inertia of the minute hands. With every tick, the hands would have to start from rest, advance a fraction around the dial, and stop. The clock simply did not have the power to drive very heavy hands. New minute hands were installed that were much lighter and, hence, had a much lower moment of inertia.

The *Horological Journal* for June 1859 reported that on Monday, 30 May the clock was connected to the west and north dials, but quarters and hours were not sounded. It was given that only two dials were in use because the hands were too heavy and these would need replacing before all four dials could be used. The report concluded that it would be some time before the *machinery could be attached to the bells*. In his 4th edition of *A Rudimentary Treatise on Clocks, Watches, and Bells*, Denison wrote, '*I then told the Chief Commissioner, for the information of the parliamentary interrogators, that the clock would be going by the meeting of the new parliament on 31 May. And so it was.*' This reference has given rise to the *official* starting date of the clock being given as 31 May, but the 30th is much more likely, since Denison was a cautious man and not one for the pomp and circumstance of starting the clock on an important day.

In early August, the clock stopped, '*Gone on Strike*', as a Times correspondent wrote. The reply in the newspaper the next day was from

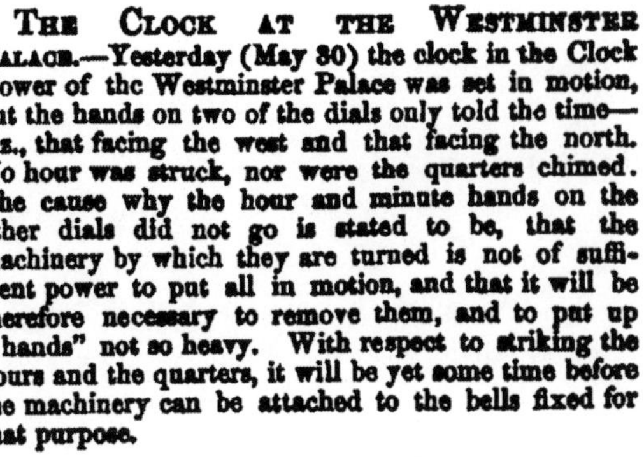

THE CLOCK AT THE WESTMINSTER PALACE.—Yesterday (May 30) the clock in the Clock Tower of the Westminster Palace was set in motion, but the hands on two of the dials only told the time— viz., that facing the west and that facing the north. No hour was struck, nor were the quarters chimed. The cause why the hour and minute hands on the other dials did not go is stated to be, that the machinery by which they are turned is not of sufficient power to put all in motion, and that it will be therefore necessary to remove them, and to put up "hands" not so heavy. With respect to striking the hours and the quarters, it will be yet some time before the machinery can be attached to the bells fixed for that purpose.

FIG 11.1 The announcement in the *Horological Journal*.

none other than the clock himself...Doubtless, Mr Denison was the author; he said that Barry's hands were too heavy and too weak, so they bent, were catching on the dial, and had to be replaced. Indeed, Denison was pilloried about the cost of the clock until he pointed out that the figure submitted to Parliament in a return included a lot of extra work that Barry had carried out. Barry was back to say that Denison had approved the hands.

THE BELLS RING OUT

The satirical magazine *Punch* reported, in its May issue, '*The most perfect specimen of dumb bells in the world are those suspended in the Westminster clock.*' Under the spoof title of '*Parliamentary Minutes*' it commented, '*That idle clock at Westminster, which may well hold its hands to its face for very shame, has cost the nation the pretty little sum of £22,057. We never knew a richer illustration of the homely truth, which is always being dinned in our ears, that "Time is money!"'*

In a letter to *The Times* dated Tuesday, 12 July, Thomas Walesby complained that Big Ben sounded poor; presumably the bell had started

striking on the Tuesday when he wrote the letter. A question had been asked of the Commissioners in the House of Commons on Monday the 11th as to why only two dials were working: Mr Fitzroy replied that new lighter hands were being made. However, no mention was made of the bell striking and this confirms the date of 12 July for the first sounding of Big Ben by the Great Clock. Letters to *The Times* soon followed, saying that the bell had been heard in Richmond Park and Campden Hill, in Kensington, but there was also a report of the bell hammer having being heard to chatter on the bell.

According to *The Times*, the Great Clock started to strike the quarters on the morning of Wednesday, 7 September 1859. Their tone was described as *extremely dull* and this was attributed to the scaffolding that was still in the belfry. The article went on to say that men in a box suspended from a derrick were working on the eastern dial and had succeeded in fixing the hands by 5 o'clock. It was expected that the other dial would be completed on the 8th. Later the *Horological Journal* for October recounted the same story but stated that the quarter bells started chiming the date on the morning of Wednesday, 14 September. This appears to be an incorrect rendering of *The Times* report.

Unfortunately the Big Ben's note did not meet with the approval of some Members of Parliament. *Hansard* reported a debate on 15 July 1859, when Mr Alderman Salomons said,

He would beg leave to ask, to whom they were indebted for the funeral notes which every hour struck upon the ear of the House? He hoped that the First Commissioner of Works, or Mr Denison, or Sir Charles Barry, or whoever it was that was responsible, would try to make some alteration in the tone of the bell. It was too bad that the Members of that House and the people should be condemned from hour to hour to hear that dreadful noise, a noise which they could only expect to hear when the great bell of St Paul's was tolled on the death of a member of the Royal Family.

Mr Fitzroy said,

He believed the bell was constructed with the greatest possible care by a gentleman who was supposed to understand the manufacture of bells better than any man in England. The combination of the metal was such as was calculated to produce the most harmonious tones. He was not a judge as to whether it had had that effect, but if the sound were an infliction, he was afraid they were likely to remain under it for a considerable time.

Mr Hankey asked,

Whether there was any chance of the bell sounding more like ordinary bells. At present it inflicted great annoyance upon the public and the House. He wished to know who was responsible?

Sir John Parkington commented that:

He thought there was no hope that the bell would ever give forth any other sound, but he would suggest that a compromise should be made. It was said the other day that two faces of the clock would not go. Why should not an arrangement be made, that all the faces of the clock should tell the hour, and the horrible tolling should cease?

If we have to take any date as the birthday of the clock, then probably it should be 7 September 1859, since it was then that the clock was striking the quarters and hours, and telling the time on the exposed dials. In a speech given to commemorate the 21st birthday of the clock, Denison gave the date of the clock's birth as 25 February 1852, the date the Commissioners wrote to Mr Dent accepting his offer and giving him authority to proceed with the making of the Great Clock.

Just over three weeks after the whole clock was operational, on Saturday, 1 October, Big Ben was taken out of service, having being found to be cracked.

THE GREAT CLOCK IS TIMED

At the end of July 1859, E.T. Loseby, an eminent London chronometer maker, wrote the first of four letters to the *Mechanics' Magazine*, attacking Denison concerning the clock. Regarding the clock, he stated that the clock did not match the original specification. First, it called for gunmetal wheels but Dent had used cast iron and obtained more money because cast iron was more difficult to work than gunmetal. However, Dent had all the wheels made with all the teeth cast in and not machined. Second, the escapement was to be a deadbeat but Denison had installed a gravity escapement, which Loseby claimed would trip. Third, the hands were required to have a visible motion, but no remontoire was installed. Fourth, the clock should be accurate to a second of time by the first blow of the hour bell, but it was not. And finally, the clock was to run for eight days on one winding and it only ran for four. Loseby also complained about Dent being paid for the clock before the clock was installed in the finished tower.

Loseby was correct in his comments, apart from the rate of the clock, but the various changes were part of the development of the clock and not unreasonable. Since Dent was not responsible for the delay in building the tower, it was reasonable that he should receive some payment for his labours. Denison never responded to these letters, having dismissed Loseby's writings a few years earlier in the same magazine as trying to settle a grudge because Denison had, '*Exposed some of his horological pretensions.*'

At the end of September, Loseby had one of his regulators installed at the Palace and Mr Quarm, the Clerk of Works, recorded the rate of the clock for several days. A long series of readings would

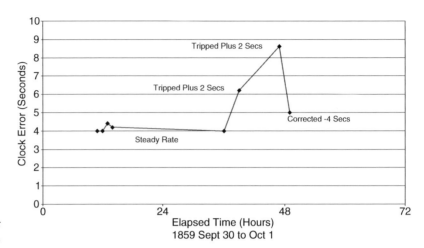

FIG II.2 Clock rate for
two days.

have been highly instructive, but the cracking of the Big Ben curtailed the readings to a few days. From his data, which were published, Loseby claimed that the clock was a bad timekeeper, and indeed it certainly appeared to be erratic. Loseby did, however, forget that it was the first blow of the hour that was required to be accurate and he drew his conclusions from readings taken at the quarter hours, where a precise let-off system was not used. Had Loseby thought more carefully and considered the data taken at the hours alone he would have seen that the Great Clock was in fact a good timekeeper; but the escapement did trip, as he had claimed that it would.

If only the hour recordings of Loseby's figures are used, an interesting result can be seen; the clock keeps good time for a day, then suddenly gains another two seconds and, after a few hours, again gains two seconds. Then the clock loses four seconds. It would seem that the escapement tripped twice, causing a total of four seconds gain. The clock was then held up by four seconds, presumably by Dent's man to correct the problem. Loseby was so close to proving his point but he failed to interpret his data properly.

We know for certain that the weather was bad—wet and windy with wind-driven sleet— when the Great Bell was discovered to be cracked. Denison wrote that the three-legged escapement was prone to tripping, and this certainly seems to have been the case at the end of September.

CHAPTER TWELVE
BIG BEN CRACKS: 1859 TO 1863

BIG BEN CRACKED

On 30 September 1859, Big Ben was found to be cracked and this was reported in *The Times* for 3 October:

The 30th was a wild and stormy day. The wind was rough in any place but round the summit of the clock tower it rushed and whistled driving the clouds of sleet through the gilded apertures till the rain trickled down in little streams from Ben and his four assistants. In the afternoon, Mr Hart, one of the gentlemen connected with the works of the clock tower, was in the belfry when the hour struck. Looking in a moment towards Big Ben he was at once surprised to perceive a row of minute bubbles spring from the bell's side at each stroke of the hammer. An instant's investigation was sufficient to show that these bubbles arose from the vibration of air in the minute cracks. Two cracks were found within two feet of each other, and opposite where the hammer struck. One was 15 inches long and the other two feet, both extending from the lip to the soundbow. Neither crack went through the bell. Striking was discontinued and Denison informed.

Denison was staying at Ben Rhydding near Ilkley in Yorkshire at the time and he soon ascertained that the bell had holes that had been stopped up with some foreign material.

THE ARGUMENTS

Dent's men had been all over the bell with a fine toothcomb. What they found were two cracks, and near to these cracks were small holes in the casting, some of which were about a quarter of an inch in diameter. A probe showed them to be as deep as half an inch. (One of Warner's quarter bells, the third, was rejected and had to be recast because it had surface defects smaller than those found on the Great Bell.) Worse than the holes, it was found that they had been plugged with some material; this, coupled with the fact that the bell had originally been painted over with some coloured wash, led to the conclusion that Mears had tried to disguise the defects. Denison made these accusations clearly in a letter to *The Times* that was published on 5 December. Charles and George Mears strenuously denied that the bell was unsound and that they had tried to disguise this.

The Times barred Denison from writing letters until such time he wrote things that were plain

FIG 12.1 The two cracks in the great bell: the closer one in the bright sunlight is the longer, the further one is the deep crack. This figure is reproduced in colour in the colour plate section.

and reasonable and not abusive. On 10 December, Denison retracted his statement that the bell cracking was caused by surface defects and went on to say that the wash that made seeing such defects difficult was put on at the suggestion of someone employed at Westminster Palace and not by Mears with the intention of concealing the holes. He said that the cause of the cracks was by a defect unknown. However, the retraction was rather half-hearted and spurred Mr Mears to action.

LIBEL

Not surprisingly, Mears sued Denison for libel; the Commissioners would only allow Denison up the tower accompanied by a member of their staff and they suggested that he bring his solicitor with him.

The libel case was heard on 30 December 1859. Denison did not enter any pleas and judgement for the case was to go by default—the defendant did not appear in court and thus admitted liability.

The counsel submitted their arguments, Mr Bovill for Mr Mears claiming that his client's reputation had been damaged as a result of the letter in *The Times* that was written by Denison. Mr Knowles responded for Denison and proposed that the counsel might come to an agreement. After discussions of three-quarters of an hour, a settlement was reached. The jury was dismissed and Denison had to retract all of his charges and had to pay all of Mears' legal and incurred expenses. (This was revealed by Mr Stainbank, successor to Mears, to be £800.) Thus Mears was vindicated and Denison had to eat humble pie.

From the hearing, it emerged that Denison had seen the bell in the foundry, after it had been cast and before it had been fettled- that is had all of the rough bits smoothed off. Therefore, he should have seen the holes alluded to. Next, it was discovered that the Palace had had the bell washed over with a bronze colour to make it look more attractive; they had also filled some holes in the bell with tin. Barry certified that the largest hole was caused by the lifting tackle pressing on the surface of the bell and opening up a blow hole just below the surface; the contractor responsible for raising the bell had filled the hole with molten zinc. Possibly the other holes had shown up after the bell had been subjected to the stress of being lifted up the tower. Considering the size of the casting and its thickness, the holes were essentially minor surface defects of no consequence.

The Times reported the court case in great detail; the article contains a glint of delight in seeing the arrogant self-opinionated Denison defeated

BIG BEN ANALYSED

The Commissioners were at a loss as to what to do with the cracked bell. Denison certainly would not have an independent view of the bell and the other referee, the Rev. Taylor, FRS, seems to have vanished. Airy was to produce a report on the clock and bells as part of his involvement with the Great Clock, so the Commissioners sought further advice from two scientists.

John Tyndall, FRS

John Tyndall, FRS, was a leading scientist of the 19th century. Born in Ireland in 1820, he was employed at first by the Irish Ordnance Survey. He studied in Marburg, Germany, and in 1853 was appointed Professor of Natural Philosophy at the Royal Institution. Tyndall was the first person to identify the greenhouse effect of the earth's atmosphere and to explain that the blue colour of the sky was a result of the scattering of light by particles in the air. Tyndall also acted as a civic scientist or government consultant and was involved in investigations of accidents in coalmines and causes of boiler explosions in steam engines. It was not surprising that a man of such standing was asked to report on the cracked Big Ben.

By December 1859, Tyndall had submitted his report on the Bell. He detailed the size and number of cavities in the bell's surface; most were small, about two tenths of an inch in diameter and about the same in depth. A few were larger, one being 5/8 inch in diameter and filled with a foreign metal. The opinion was that they were caused by air bubbles trapped by the slope of the soundbow. Two cracks were reported. One crack was exactly opposite where the hammer struck and thus was at a point of maximum vibration. The other crack was about 15 inches from the first. Three other small cracks, only two inches long, were recorded. However, no sign of these can be seen today, so these were probably just small surface defects that can occur when the loam mould cracked slightly. Tyndall concluded that the cracks were not caused by the cavities, but instead probably a consequence of the brittleness of the metal and the size of the blows of the hammer. He suggested that the position of the bell hammer be changed to produce a different pattern of vibrations and the crack monitored. If the crack increased, then the bell would have to be recast.

John Percy

John Percy was the first Professor of Metallurgy in the Royal School of Mines in London. He was born in 1817 and trained as a medical doctor, but he turned to metallurgy after inventing a process for extracting silver from its ores. His book, *Metallurgy*, spanned five volumes but was not finished before he died. As the leading metallurgist of the time, it was appropriate that he be asked to analyse the metal of big Ben.

In March 1860, Percy reported his findings. Denison had specified that the composition of the bell should be, by weight, 7 parts of tin to 22 of copper. The alloy used by bell founders was 7 parts of tin to 24.88 of copper. George Mears of the Whitechapel bell foundry endeavoured to persuade Denison that the alloy he had chosen was too hard, but Denison refused to be convinced and Mears was bound by his contract to use the harder alloy. Mears concluded a letter of 9 May by stating,

Mr Denison's alloy, as before stated, of 7 of tin to 22 of copper, is a much more brittle material, and was adopted by him against my remonstrance; and in addition thereto the bell was struck by a hammer of unprecedented weight, so that I assert not a shadow of blame can attach to me for the result.

Percy had samples from mouldings at the top and bottom of the bell analysed; one was close to Denison's specification for the bell composition; the second, taken from the soundbow, contained an additional two per cent of tin. Percy acknowledged that this was not necessarily a true representation of the overall composition of the bell. Percy stated that the proportion of tin in the Westminster bell was excessive, causing brittleness. Mears, on commenting on the variation of tin content, wrote:

In reference to the difference of composition between the upper part and lower part of the bell as found by Dr Percy, I beg to state that it is a well-known fact that the alloy of tin and copper is an unstable one, which no skill in founding has

FIG 12.2 Dr Percy's excavation to determine the depth of the crack and to remove samples for analysis. A cluster of holes can be seen at the top of the larger excavation.

ever yet overcome, and the result of Dr Percy's examination does not give greater variation than I have myself found.

The Office of Works then asked Percy, in cooperation with Mears, to drill out more metal and perform a more meaningful analysis. Samples were taken from the main crack, chipping down the side until the bottom of the crack was reached. It was 3 inches deep. Today a hole in the bell the size of a tea cup can be seen where Percy took samples for this work. The chippings are today preserved in glass jars in the Science

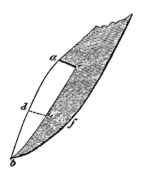

FIG 12.3 Diagram produced by Dr Percy to show the extent of the principal crack, which extends to the lip of the bell.

Museum. Percy reported back in May of that year; analysis of the removed metal showed that in the region of the crack, the bell's composition contained about 2% more tin that had been specified by Denison. Percy's conclusions were that the bell was too brittle and that the hammer was too heavy, and that this combination had caused the bell to crack.

Denison lifted some unconnected words from Percy's report, which he used from time to time. They were '*porous*', and '*inhomogeneous*.'

AIRY'S REPORT

Airy reported on both the clock and the bells in April 1860. Regarding the clock, he suggested that Denison's new gravity escapement be fitted. Airy concluded that the crack did not affect the tone of the bell and that the hammer was unreasonably heavy. Since the crack was at the point of maximum vibration, he proposed that the bell be rotated by an eighth of a turn so that the principal crack would be at a node of minimum vibration. Upon reducing the hammer weight and turning the bell, Airy recommended the striking be resumed and the crack checked on a weekly basis. He also suggested that the hammer head be faced with tin to give a softer blow.

Denison wrote to the Commissioner on 3 April, saying of Airy's report, 'But this is not the only instance I have seen, professionally or otherwise, of Mr Airy committing himself to hasty and erroneous conclusions on subjects of which he has no practical knowledge.' Denison was using his standard defence technique of attacking his perceived opponent.

Another letter from Denison followed on 11 April 1860; this time discussing a new bell and how it might be made lighter. There must have been some contact from Cowper, the First Commissioner, to precipitate such a communication.

Denison was asked what work was outstanding on the clock; his answer of 25 April 1860 reported: 'Finishing the escapement that was left to last that we might see the effect of the wind on the new hands and through them upon the clock— and some apparatus for stopping the winding whenever the weights are wound up.' Denison also mentioned that, 'The reason the time is not being shown on the south dial is that Dent's men found Sir C. Barry's hour hand cracked when they were putting on the new minute hand.' Never one to miss an opportunity, Denison went on to attack Barry's design for the hands and to remind the Commissioner that Denison's minute hands were only a third of the weight of Barry's.

It certainly appears that Barry had been advised by Vulliamy on making hands out of a casting of gunmetal. Many of Vulliamy's turret clocks used cast gunmetal hands that are so delicate it is difficult to see today how such a thin casting could ever have been made.

FIG 12.5 The largest blowhole in the casting and a surface mark caused by the mould surface cracking. Neither are of any significance on a casting of this size.

An interesting insight into the state of the clock was given on 11 June 1860 in the House of Lords, when The Earl of Derby observed. '*One of the hands had disappeared altogether, and the other stood at twelve, so that it had the merit of being right*

FIG 12.4 At 6½ cwt, the original hammer head is a massive piece of cast iron. It is now on display in one of the tower rooms.

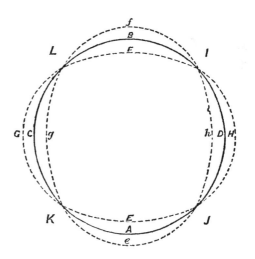

FIG 12.6 Airy's diagram to explain the bell's vibration, which is at a maximum (A–D) where the hammer strikes, directly opposite, and at 90° to these points. Nodes of minimum vibration (I–L) are at 45° from the maximum.

at least once in the twelve hours.' He went on to say that he, '*in common with all the inhabitants of that part of London in which he lived, rejoiced that the great bell had been cracked, and he trusted that no attempt would be made to make the clock speak to their ears again in the old tones. He should like to see the hands in motion again, but he hoped the bell would remain mute.*'

On 16 June 1860, more questions were asked in the House about the stopped clock; this was reported to be for alterations to the clock and the bells. This is the most likely date that the new double three-legged gravity escapement was installed.

WARNER'S LETTER

John Warner and Sons cast a new 81 cwt bell 'Victoria' for Leeds Town Hall and on 12 January 1860 a letter of his appeared correcting some details in a report of the previous day. With the cracking of Big Ben the second two months earlier, Warner must have enjoyed the opportunity to state his case as to why the first bell failed; absolving himself from responsibility and placing the whole blame squarely on Denison's shoulders. He wrote,

The subject of the proper mix of metal for large bells has been brought recently so prominently before the public in connection with the unfortunate Big Ben the second, and to state that we were compelled, under direction, to construct the original Big Ben of the former proportions—viz., 7 to 22, or nearly 1 part to 3—we have never adopted that mixture for any bells the construction of which was left in our hands, and had the original Big Ben been formed of the usual mixture, and been struck with a clapper weighing from 5 to 6 cwt, instead of one of 13 cwt, and had it not been allowed to come in contact while in a state of vibration from the action of the clapper with the ponderous experimental clock hammer fixed on the

outside, the probability is that the second bell never would have been required.

EXCHANGES IN PARLIAMENT

The First Commissioner, Mr Cowper, answered a question in the House of Commons about the bell on 8 March 1860, saying, '*The Great Bell in the Clock Tower was in the unhappy condition of being cracked in five places—the first crack was twelve inches, and the second nine inches in length. He was not able to state the origin of the cracks, but thought he was safe in saying that either the hammer was too heavy for the tenacity of the Bell, or the Bell was too brittle for the weight of the hammer.*'

On 4 June 1860, Mr Cowper again addressed the House of Commons. 'He considered that the best course to be taken as a temporary arrangement was to use the largest quarter bell for striking the hour, and the other three quarter bells for striking the quarters. If that were done, the House would be spared the loud and dismal strokes of the great bell, which they must all remember, and which occasionally drowned the voices of Hon. Members, and diverted attention from the business before the House.'

The sentiment against the Great Bell was echoed in the House of Commons by Sir George Boyer, who said on 3 August 1860: '*It had been a perfect nuisance to the whole Metropolis, and he hoped would never be used again for striking the hours.*'

These and other references indicated that the note of Big Ben was loud and the striking was ponderously slow. No further information about the speed of striking has been uncovered but it seems likely that it was speeded up. Most probably, the fly blades were made smaller.

A TEMPORARY SOLUTION

A proper solution to the cracked bell was to take almost four years to effect, so in the meantime a temporary plan was devised. In June 1860, the Astronomer Royal had proposed that the hours be struck the on the fourth quarter bell. James Turle, the organist at Westminster Abbey was requested by the Commissioners to comment on the bell. He replied on 11 September 1860, and suggested that the third quarter bell be substituted for Big Ben, to sound the hours. His advice was not taken but the fourth quarter bell was chosen instead to strike the hours. This meant adding another hammer to this bell; its third, since two hammers were needed to sound the last quarter. In addition, extra levers were needed to convey the action from the hour-striking train to the fourth bell. Jabez James, an engineer who did much work about the Palace, started the alteration on the last day of 1860 and it appears that the work was completed in February 1861. Shortly after the hour striking was reinstated, a brief article in *The Times* reported that the Great Clock had struck the wrong hours on 1 April 1861; the writer impishly suggested the Clock was playing an April fool's trick.

When it next sat in February 1862, Parliament asked no questions about the Great Bell.

In July of that year, Jabez James submitted a bill for £56 17s. 3½d. This included £25 for altering the chimes and for work on the hands, and a fee of £5 5s. 0d. to James Turle.

The music must have been rather unsatisfactory, lacking the full majesty of the hour bell, but the interim striking system bought time for the proper solution to be devised. It also allowed memories to fade, so no one at all got blamed and Denison escaped, since he was an unpaid adviser to the Government.

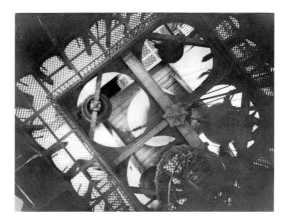

FIG 12.7 A view of the bells taken by the press photographer Hodsall around 1907. The fourth quarter bell is in the lower right-hand corner and the third hammer can be seen pointing towards Big Ben. From the London Metropolitan Archives.

Today there are some surviving relics of the third hammer. In the link room there is an unused bell crank that was once part of the connection to the fourth bell, and underneath the wooden staging below Big Ben is the third hammer, which was removed from the quarter bell some time after 1907.

FIG 12.8 Now stored under Big Ben: the hammer that once sounded the hours on the fourth quarter bell.

BEN STRIKES AGAIN

In February 1862, the Whitechapel Bell Foundry quoted for the required work of turning Big Ben so the cracks were at a position of minimum vibration, fitting a lighter hammer head to replace the one weighing a heavy 6½ cwt, and installing some wooden staging just below the bell. In the unlikely event of Big Ben cracking catastrophically, the wooden platform would catch the parts of the bell as it fell, thus, preventing it from falling through the floor onto the clock.

Mr Cowper's final address on the bell was on 10 April 1862, in which he said, 'The disordered and cracked state of the bell had given him great anxiety. He had called in the most scientific advice he could find. There had been a consultation of learned doctors, and he had already laid on the table four Reports.' He mentioned that the bell was too brittle and concluded by saying that, 'He apprehended there was no wish on the part of Hon. Members for a repetition of the loud and slow sounds of which they complained before. Under these circumstances, he had not come to any conclusion as to what should be done with the Great Bell. At present the larger quarter bell was used to strike the hours, and though that arrangement was not perfect in regard to musical harmony, he thought it convenient for the present.'

No action was taken by the Commissioners until the end of 1862, when they wrote to the Whitechapel Bell Foundry asking them to examine the bell to see if it could be sounded again less loudly and less slowly, without altering the bell's quality, note, and tone. On 11 January 1863 the foundry reported that on hearing the Great Bell being struck, the tone was *uninjured*. A bell is a complex musical instrument that produces several distinct notes (harmonics), each note coming from a different part of the bell. When all these notes are in tune with each other, as produced by the bell founder's art, the characteristic pleasing bell note is obtained.

Big Ben has a deep beating note that is quite noticeable: in fact it is this beat that helps give a distinctive character to the sound of Big Ben. Technically, the beating is known a doublet and on Big Ben, this occurs at a frequency of 3.7 Hz. Doublets in bells are well-known and are produced when the bell is not exactly symmetric about the vertical axis. Another cause in the case of Big Ben could be the cracking and the removal of metal to determine the depth of the crack. Bill Hibberts analysed the notes of eight other large bells cast by Mears in the 19th century. The largest was Great Tom of Lincoln, cast in 1835, and weighing about 5¼ tons. Typically, these other Mears bells did not have a significant doublet, as displayed by Big Ben. This implies that Mears were capable, as one would expect them to be, of casting large bells with a sufficient degree of axial symmetry.

It is therefore probable, but not certain, that the beating in Big Ben's note is as a result of the two cracks and the excavation, rather than a casting defect. Analysis spent with the bell itself rather than a recording, tapping around the rim to see if the degree of beating is linked to the position of the crack, might add evidence one way or the other.

One might imagine that Denison would not have approved of such a tone in a new bell, but it is quite probable that his musical ear was not very good. Language changes, but the use of the word *uninjured* gives a different meaning than *unchanged*. Perhaps the bell founder, and indeed all the parties involved, wanted Big Ben ringing again as soon as possible and a slight change in the character of the note was a small price to pay.

At last, on 29 August 1863, the Commissioners wrote to the Whitechapel Bell Foundry, giving them the order to proceed with work on the Great Bell; to provide a new lighter hammer head, turn the bell as per Airy's direction, and to construct a wooden platform under the bell. The whole work was to cost a maximum of £220.

On 14 July 1863, there was a reference in *The Times* to the bell still being cracked; one wonders if the newspaper had an idea of what was going on or if the Office of Works responded to the report. A poem in *Punch* in July 1865 speaks of Big Ben striking. A comment in *The Times* of 25 November 1864 lists a great many Government mismanaged projects, but the Great Bell is not mentioned.

It was not until Thursday, 5 November 1863 that Big Ben was again heard striking. Presumably, it was viewed as a non-event by the media; *The Times* made no comment and it fell to *Trewman's Exeter Flying Post* or *The Plymouth and Cornish Advertiser*, through their 'London Correspondent' to inform the public as follows:

This Thursday (5 November) we were startled by the tolling of a huge bell: every minute or thereabouts a deep sepulchral sound rose high in the air, booming away in very unearthly music. What was it? Could it be the bell of St Paul's announcing the sudden death of a member of the Royal Family? Nobody knew what it meant, until by slow degrees it dawned upon our minds as well as our ears that Big Ben had again found his tongue. And such was the fact. Since his fatal crack the hours have been struck upon one of the quarter bells, until last week, when in consequence of a successful operation he recovered his voice. It might have been in tune in his young days, but could have never been very musical, and it is now decidedly less melodious than ever. I have not been able to find out precisely the manner of the cure, but believe it to be a mixture of stopping and boring—the latter to prevent the crack from spreading, the former to obviate the jarring on a crack. The operation has so far succeeded that Big Ben No 2 has recovered a certain amount

of vocal power, but the noise he makes is just such as much used in old Richardson's show to accompany the appearance of the ghost, but only it is a bigger and a more voluminous noise, as it ought to be considering the width of his throat.

A four-line note appeared in the *Observer* on 15 November 1863 to say that Big Ben was again announcing the time from the Clock Tower of Westminster Palace, the founders having found that the bell was not so cracked as people imagined.

Parliament was in recess when striking was restarted and when it reconvened in February 1864 there was no mention of the Great Bell, so the two houses must have been content with the solution.

John Percy, the metallurgist, inspected the Great Bell again in August 1865: he reported that he had not been able to detect any extension in the cracks. The two cracks are today still stable and have not moved; however, they are still regularly monitored by the Whitechapel Bell Foundry, which has had the care of the bells since they were installed. Apart from regular maintenance and some major overhauls on the hammers and fixings, Big Ben has been sounding the hours without problems. From his installation to the clock's 150th anniversary in 2009, Big Ben would have sounded around 8.5 million times... not bad for a cracked bell!

Years later, in, 1865 *Punch* produced some doggerel titled *Lay of the Bell*. Its second verse ran:

To me charming fellows have bowed
Their silver tones sweet as a coo
Big Ben of his notes was so proud!
And a stir made in Parliament too
But when Ben had got up in the world
ThLere was one thing—sound sense, which he lacked
And I found while his lofty lip curled
Poor Ben was a little bit cracked.

1865

In 1865, Denison had occasion to write to *The Times* reminding everyone of Percy's report that, *'The bell was porous, unsound, inhomogeneous and not of the prescribed composition.'* Denison went on to ridicule a report by Dr Tyndale that said that the bell was too brittle and the hammer too heavy, saying that, *'The Commons laughed at it.'* Airy was involved and Denison said that, *'Airy had replaced part of the bell mechanism with a contrivance of his own, but it had to be removed.'* Today, a powerful leaf spring lies unused in the striking train; perhaps this was Airy's addition.

WHY DID THE BIG BENS CRACK?

Today, there can be little doubt that Denison got it wrong. He specified a bell metal that contained too much tin, making it more brittle than conventional bell metal. He also used hammers that were excessively heavy and his treatment of the Warner's Big Ben was cruel. Warner managed to cast a bell that was too thick and it is surprising that Denison, who specified the size and shape of the bell, did not have the bell's dimensions checked carefully when the bell was so overweight. Likewise, if there were holes in Mears' bell, Denison should have found these whilst the bell was in the foundry. Reports of the striking of Big Ben in the clock tower tell of the bell hammer chattering on the bell. There was an unsatisfactory means of keeping the bell hammer off the bell. This could have been a major factor in the cracking of Big Ben II. Similar reports of hammer chattering had been made; again this could have been a contributing factor.

Denison advised on and designed bells and peals of bells. Today, all but one of his rings has been recast; campanologists have not held Denison's bells in any regard for their tone.

Perhaps the whole event of the bell cracking and its aftermath could be characterized by the following parody on the nursery rhyme *Oranges and Lemons*. It appeared in *Punch* magazine on 24 December 1859 after the bell had cracked and seems to reflect the public exasperation with the whole clock tower saga.

The Bells of Big Ben

Big Ben's case looks scaly, say the bells of Old Bailey;
His voice is quite gone, say the bells of St. John;
He's chock full of holes, peal the bells of All Souls;
Must go to the forge, chime the bells of St. George;
Even my voice is sweeter; sneer the bells of St. Peter;
He ain't worth two fardings, snarl the bells of St. Martin's;
Case of too many cooks, say, growl the bells of St. Luke's;
Don't know what they're about, howls St. Botolph Without;
MEARS, DENISON chides, say the bells of St. Brides;
Well, d'ye think MEARS is wrong? asks St. Mary's ding-dong;
I don't, if you do, says the belfry at Kew;
It's a great waste of tin, tolls St. Botolph Within;
And the cash must come from us, growl the bells of St. Thomas;
Aye, every shilling, add the bells of St. Helen;
And we're not over-rich, groan the bells of Shoreditch;
It makes one feel ranc'rous, say the bells of St. Pancras;
Yes that's for sartin, again rings St. Martin;
But what's to be done, once more peals St. John;
Bang'd if I know, tolls the big bell of Bow.

CHAPTER THIRTEEN
THE CLOCK TICKS ON: 1863 TO 1900

PERFORMANCE

In November 1875, the *Horological Journal* reported that the error of the Great Clock was less than 1 second on 83% of the days of the year, as reported by the Astronomer Royal. Since the dial could not be read with great accuracy, it was the first stroke of the bell that signified the hour. Since the speed of sound would cause a delay in observing the time that the bell sounded, a map was appended that showed the allowance that had to be made to obtain the correct time. People in Clerkenwell would have to allow 12 seconds for the sound to travel from Westminster, whereas if the bell was heard in Greenwich Observatory, the delay was 26 seconds. For the very precise-minded, it was noted that it took a quarter of a second for the sound to travel from the bell in the tower to ground level at the tower base.

1863

In June, *The Times* announced that, in his annual report, the Astronomer Royal recorded that the Great Clock varied by less than second a week, much better than the second a day originally specified.

1864

As building of the Houses of Parliament progressed, houses and buildings in and around New Palace Yard were demolished. Fendall's Hotel and another building stood at the base of the clock tower. In *The Times* a letter asked when was the demolition of these buildings, which had been going on for two months, was to be completed? Similar questions were asked about non-completion of Government projects: Nelson's Column, the New Law Courts, and, indeed of the Houses of Parliament.

1865

Denison also presented a lecture on large clocks to the British Horological Institute; an event that might have won him the invitation to be their President some years later.

1868

The British Horological Institute was founded in 1853 as a trade body to provide education for the clock and watch trades. In 1868, the Institution invited Edmund Beckett Denison to be their President. He accepted and held the post

1. The Belfry : Big Ben and His Little Brothers.—2. The Pendulum.—3. Behind the Dial.—4. Winding up the Machinery.
INTERIOR OF THE WESTMINSTER CLOCK TOWER

FIG **13.1** A montage view from *The Graphic* of March 1876.

until he died in 1905. Edmund was a faithful president but caused a major upset during a prize-giving in 1873. He said that the prizes were given for mere finger work and polishing. '*Working for fools*' was an expression he used, where makers polished parts that were not working surfaces.

1870

Coconut fibre was used at the base of the weight shaft to cushion the fall of the weights should a weight line ever break. Immediately beneath the clock, the top of the weight shaft was protected by an iron grating and, to prevent draughts, this

174

grating would have been covered with matting or a floorcloth. In the event of a fire, this would have given no protection at all.

On the morning of 21 June 1870, the coconut matting caught fire, ignited from a furnace in the basement that was probably part of the ventilation system. Four tons of matting were badly damaged but the fire was quickly spotted and brought under control. Had the fire taken hold, the weight shaft would have acted as a massive chimney and the clock at the top would probably have been so badly damaged that it would had to be replaced. In the event, the clock only needed to be stopped for a few days, to clean it following smoke damage from the fire.

1871: THE AYRTON LIGHT

The Ayrton Light is a very important feature of the Clock Tower and has a passing connection with the clock dials. On 7 June 1871, a paragraph in *The Times* entitled '*Notice to Truant MPs*' announced that the First Commissioner for Works intended to install a light in the top of the Clock Tower to be lit when the House of Commons was sitting. The light was to be installed under the direction of Drs Percy and Tyndall. On the 27th of that month, a question was asked in the House of Commons as to what was happening. Ayrton, the First Commissioner, replied much along the lines of *The Times* article but added that an experiment might be made first with a light on the Victoria Tower.

At first, lime light was mentioned in *The Times*. This employs a gas jet to heat a cone of lime to such a high temperature that it becomes incandescent. Very soon afterwards, it was announced that a magneto-electric light was to be used, since it was more powerful. By March

1872, little had happened and a question in the House of Commons revealed from Mr Ayrton that, '*The proposal to exhibit the electric light on the Tower while the House of Commons was sitting was very fully considered, and it was found that to erect and maintain the light would involve an expense larger, perhaps, than the circumstances would justify.*'

Shortly afterwards, in June 1872, there was a fire in the Clock Tower caused by the light igniting a wooden screen but no damage was caused. It is worth noting that wood is only used in the Clock Tower for doors, the ceiling of the clock room, and the platform under the bell Big Ben; the tower is essentially a fireproof building with a roof of cast iron tiles and supporting beams of wrought iron. In exchanges in the House of Commons, a suggestion had been made to use the clock dial lights instead of a dedicated light but the reply was, '*As to substituting the light behind the face of the clock for the light used in the tower, great inconvenience would arise from extinguishing the light behind the clock face at uncertain times.*' This rather implies that the gas jets behind the dials were lit and extinguished manually.

In April 1873, experiments were carried out to determine the best form of light in the Clock Tower. *The Illustrated London News* of 11 August 1873 carried a report of one experiment and engravings of the lighting apparatus. Essentially, this consisted of a Gramme electric generator situated in the basement of the House of Lords and driving a lamp system in the Clock Tower. An arc lamp was placed at the focus of a lighthouse-type lens, as the carbon burnt out, a second lamp was brought into action and moved into the position of the focus. An obituary in the *Horological Journal* of February 1929 recorded the passing of the engineer Conrad W. Cooke, who was reported to be the first man to light Big Ben

ELECTRICAL APPARATUS FOR THE CLOCK-TOWER LIGHT, HOUSES OF PARLIAMENT.

LIGHT ON THE CLOCK TOWER, HOUSES OF PARLIAMENT.

FIG 13.2 Experiment with an arc light in the Clock Tower.

with electricity and to shine a beam of light over London.

A question in the House of Commons on 21 April 1874 asked if the experimental light was to be retained and Commissioner replied that this was up to the Members to decide. His answer did, however, indicate that the temporary light on the Victoria Tower was still in position. By June 1874, it seems that no one objected so the light was to remain. An engraving of 1877 shows the light, which certainly appears to be a gas light, probably employing Argand burners.

The light, which proclaims that the House of Commons is sitting, became known as the Ayrton light, after Acton Smee Ayrton, who was the First Commissioner for Works when the light was installed. The light is still used today, although of, course, it is now electric.

THE BEACON LIGHT IN THE CLOCK TOWER

FIG 13.3 The Ayrton Light in the Clock Tower as depicted in *The Graphic* of 29 October 1887.

No. 51.   STATESMEN, No. 34.
"Mind and Morality."

FIG 13.4 A *Vanity Fair* cartoon of Acton Smee Ayrton by 'Ape'. The title is *'Mind and Morality'* indicating his Parliamentary principles. This was published in October 1869, before the Ayrton light was installed.

1873

The going train was stopped for a few days, so that it could be cleaned.

1877

The going train was stopped for three weeks so that a scaffold could be erected in the clock room for repairs and redecoration. Repairs were to be made to the clock and a new escape wheel fitted. A report said that, '*The old escape wheel was taken away to be repaired before the clock was stopped, but by the aid of a number of workmen it was kept going as well as if it had been controlled by its pendulum.*' This bizarre situation could only mean that the workmen were manually letting the clock run at a controlled rate, no doubt under the supervision of a chronometer. That a new escape wheel was needed after around 17 years of service on the surface is, at first glance, a mystery. However, although the escapement was only 17 years old, it was probably at this time that it was discovered that the gravity escapement had an innate ability to knock itself to pieces. With a solid tick every two seconds, a good escapement must have the double three-leg escape wheel fixed to the arbor on a hexagonal or octagonal spline. Merely to pin the collet of the escape wheel means that, in time, the holes for the pin get worn into an oval and the escape wheel becomes loose. Likewise, the lifting pins must be securely riveted or screwed into the escape wheel, such that the rubbing motion tends to screw the pins in tighter. Finally, the locking blocks on the gravity arms must be pinned steady as well as screwed to withstand the relentless knocking as the escape wheel locks. Denison later wrote about the need to secure the escape wheel properly. Most probably, the new escape wheel was made to resist these previously unforeseen issues and a spare was made, or the old one repaired and kept for emergencies. It might well have been a wheel made in 1877 that was the spare used in 1976 when the clock was set going so quickly after the disaster, since a spare escapement was available.

Regilding of the spire and external decoration was carried out. A time capsule was filled with coins and newspapers and duly sealed and deposited in the spire.

1878

Sir Edmund Beckett

The Times published a letter on 31 October 1878, from Sir Edmund Beckett. Edmund Beckett Denison had inherited the baronetcy from his father and dropped the Denison name, but I shall continue to refer to him as Denison, for simplicity. The subject of the letter was the bells of St Paul's cathedral; in this Beckett wrote of Big Ben *'The bell was a disgrace to the Nation, as it was to its founders.'*

'The Oldest and Worst of the Foundries in England'

One might have thought that with Big Ben successfully striking the hours with no further mishap, the history of the Great Bell might have been let to lie; but it was not to be so. By the time the next battle occurred, Mr Mears had died and Mr Stainbank was running the bell foundry at Whitechapel. Stainbank joined the business in 1861, the foundry becoming Mears and Stainbank in 1865.

On 20 November 1878, another letter from Denison appeared concerning the bells of St Paul's, telling of his new method of suspending bells, which was employed by all the bell founders except *'The oldest and worst of the foundries in England.'* This was an allusion to the Whitechapel foundry. The next letter from Denison was published on 17 January 1879; he wrote that Mr Stainbank was aggrieved at his statements about the Great Bell, adding, *'It was true that Mears got it certified by the referees by filling up the holes before we saw it, knowing perfectly well our opinion that porosity was unsoundness.'* Stainbank replied to *The Times* to assert that Denison's statement was untrue and that they would sue him. Denison responded with a fount of vitriol about Mears,

Stainbank responded by suing Denison for libel. After Denison again replied, the Editor decided to draw the correspondence to a close.

1879

On 29 June 1879, Denison found himself again in court, defending his comments on the bell Big Ben. The three letters from *The Times* were presented as evidence and it was ruled that it could not be decided by the Judge in Chambers, since an issue of law was undecided. *Had Denison's comments been directed at the late Mr Mears or at the company?* If they were directed at Mears, then there would be no case to answer, since he was dead and, in English law, you cannot libel a dead person. If the comments were directed at the company, then there was a case to answer.

The Court of Appeal heard the case on 10 December 1879. There seems to have been no outcome in the appeal and matters slept for a short time, until 1880, when Denison again penned a letter to *The Times* that appeared on 23 November; the subject again was the bells of St Paul's cathedral. In his letter, Denison said that Big Ben was *'Porous, unhomogeneous, unsound'* and that he wrote *'to prevent the referees being taken in again by concealed defects as we were in 1859.'* Mr Stainbank responded the next day to assert that Beckett's statement was untrue and that they would sue him. Denison responded a day later, saying *'Mr Stainbank's action has been many times before Masters, Judges, Divisional Courts, and the Court of Appeal… They could do no more until it comes to trial some day.'*

Dent's Drawing

Frederick Dent was succeeded by E. Dent & Co. In 1879, they commissioned a pen and wash

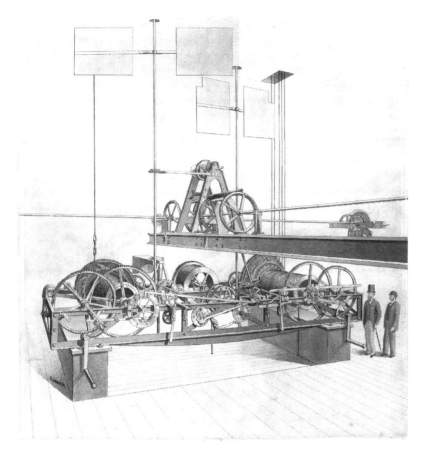

FIG **13.5** Dent's drawing of the Great Clock. This figure is reproduced in colour in the colour plate section.

drawing of the Great Clock, which was used in their publicity material. It is interesting to note that the drawing is really quite accurate, though the men on the right are out of scale; but perhaps it was drawn like that to make the clock look larger. One thing is missing though; the inscription along the bottom that records Frederick Dent as the maker and E. B. Denison as the designer. Perhaps Elizabeth's successors (she died in 1865) were still very sensitive about Denison's past involvement with the company and the dispute over Frederick's will. Also, they might have been embarrassed by the complaint of which Frederick died.

1880

The west dial needed attention and, in July 1880, £240 had to be spent in executing the necessary repairs. These involved painting and regilding the dial, and repairs to the surrounding ironwork, the hands of the clock, the metal framework, and the opal glass. These repairs were reported to have been of a dangerous nature, could only be undertaken during the summer months, and a special scaffolding had to be designed and built. This was the first time that the clock face had been cleaned since it was erected.

1881

The case of Stainbank versus Beckett was finally heard at a *nisi prius* court at Westminster on 27 June 1881 and lasted seven days. Thomas Loseby, watch and clockmaker of Leicester, testified having made a plan of the holes and cracks in Big Ben; he also took castings of the holes. William Wanskitt, who was a bell founder at the Whitechapel foundry, testified that he was there when Big Ben was cast. There were cavities in the casting but he did not consider them serious. Under cross-examination, he explained that the holes were filled with a stuffing made of zinc filings and resin; the filling being done before Beckett and the Rev. Taylor first examined the bell. Henry Hart, who was foreman to Jabez James, the contractor for the raising of the bell, said that he had been told to bronze the bell with nitric acid.

At the end of the trial, after three-quarters of an hour the jury decided that one letter was libellous towards the company, that of 20 November, in which Beckett wrote '*The oldest and worst of the foundries in England.*' The jury awarded £200 in damages to Stainbank, the plaintiff. The following day, Messrs Curtis & Betts, solicitors of Mears and Stainbank published a letter in *The Times*, clarifying the jury's ruling that all letters were libellous, but only one referred to Stainbank. Denison came back the next day to repeat that, '*Big Ben was a disgrace to its founders*', and that, '*They got paid for it by concealing its defects from the referees;*' statements carefully chosen from the letters that were not thought to be libellous to the company Mears and Stainbank. In a clever turn of words that avoided further libel, he still managed to include his libellous statement as a last stab in a skilful way that only a lawyer could concoct: '*I was not*

much surprised at a British jury, in spite of what I was told was the very favourable charge of the Judge, finding all does not justify one in calling that foundry the worst of the very few in England.' He ended by wishing Mears and Stainbank '*the joy of the result.*'

THE 1880S

Denison bought an estate near to St Albans, where he built a house of his own design in the Queen Anne style; it was named Batchwood and had a tower with a turret clock by Joyce of Whitchurch and a bell. Denison engaged in the restoration of the west end of St Albans Abbey, totally rebuilding it in a modern style very far from the original mediaeval appearance. For this work, he gave his name to the English language. The *Oxford English Dictionary* defines to 'Grimthorpe' as to '*restore (an ancient building) with lavish expenditure rather than skill and fine taste.*' However, if Grimthorpe had not stepped in and organized the restoration, it does seem rather probable that the west end would have soon collapsed. Denison also restored two churches in St Albans. Of himself, he said '*I am the only architect with whom I have not argued.*'

Denison gave advice freely on clocks and bells; mostly to churches. He also became the Chancellor of the York Diocese, a high ecclesiastical legal position. It was for his services to the Church of England that it is believed he became a peer.

1886

Sir Edmund Beckett was raised to the peerage and entered the House of Lords as Lord Grimthorpe on Tuesday, 23 February 1886. It was almost three years later that the magazine *Vanity*

FIG 13.7 A pastel portrait of Lord Grimthorpe. This figure is reproduced in colour in the colour plate section.

FIG 13.6 *Bells*: The *Vanity Fair* cartoon of Lord Grimthorpe by 'Spy' that appeared on 2 February 1889. This figure is reproduced in colour in the colour plate section.

Fair wrote a satirical review of his Lordship and the cartoon bore the title '*Bells*', no doubt a painful reminder of his defeat in the court some eight years earlier. In the accompanying text, the briefest of mention recorded that, '*The bells at Westminster owed their parentage to his versatile talent*,' perhaps a case where saying nothing was more powerful than stating the obvious. Regarding his Lordship's involvement with architecture, the text commented, '*He is so high an authority on church architecture he is credibly reputed to know the difference between a gargoyle and a flying buttress.*' The writer concluded with the dry words, '*He is not a radical.*'

CHAPTER FOURTEEN

THROUGH TWO WORLD WARS: 1900 TO 2000

LORD GRIMTHORPE

Edmund Beckett Denison, First Baron Grimthorpe, the father of Big Ben and the Great Clock, suffered a stroke in 1899. On Saturday 29 April 1905, and with his sister, Mrs Paget, at his bedside, he died from heart failure. It was reputed that late in his life he said, '*I have no friends and all my enemies are dead.*'

Edmund specified in his will:

And I desire that no flowers or other decorations be used at my funeral or on my grave and no glazed hearse or lead or other strong coffin or brick grave be used, that the grave be not deeper than four feet, and that no stone monument be put up for me as my buildings are enough.

On 3 May, Edmund's remains were interred with those of his wife in the north-eastern corner of St Albans Abbey graveyard, an area that was reserved for Bishops and the like. The dean of Canterbury read the committal and the Bishop of St Albans pronounced the blessing. Edmund got some of his requests; the hearse was a plain black one, there were no flowers, and he would have probably approved of the moss-lined grave. However, the coffin was polished oak with a large gold cross on the lid and, later, a plain red granite slab was raised to mark his resting place; the inscription records him as the restorer of the Abbey but makes no mention of his part in creating the Great Clock and Big Ben.

The funeral service in the Abbey was led by the Rev. C.V. Bicknell, vicar of St Thomas in St Albans, which was Edmund's parish church. A large and representative congregation attended the funeral. In the town, flags were at half mast, businesses suspended their activities during the service and a general air of mourning pervaded St Albans. Mourners included Edmund's two sisters, nephews and nieces, and clergy from the various churches he had supported over the years. Also there were the Mayor and Corporation of St Albans, the Lord Lieutenant of Hertfordshire, and Lord and Lady Verulam. The British Horological Institute was represented by James Haswell, Vice-President and Chairman, and Edwin Desbois, the Treasurer. Other horologists included James Hall, FRAS, and Robert Potts, of the Leeds company of turret clockmakers. In the evening a muffled peal was rung.

Edmund Beckett Denison's Character

Edmund had a most colourful character. If there was any feature that characterized him, it was his dominating attitude, where he commanded every situation. Not surprisingly, he had few friends, if any, but he was remarkably generous as well as being deeply absorbed in several subjects. This astonishing imbalance suggests

FIG 14.1 Lord Grimthorpe as an old man.

that there was some underlying issue that made him what he was.

Asperger's syndrome is a personality disorder relating to autism and was first identified by Hans Asperger in 1944. The most obvious problems that people with Asperger's syndrome might experience are language difficulties, especially caused by difficulties in understanding nuance and body language, and in taking phrases literally. They might be able to communicate as a speaker but have problems entering into a conversation that involves exchanging information and modifying and developing what they think and say. Additionally, difficulties are found in forming friendships: a lack of empathy with people and an obsessive occupation with certain

subjects can be observed. Also, people with Asperger's syndrome generally follow routines rigorously, might have strange body motions, be prone to clumsiness, display unusual facial expressions, have a formal or unusual manner of speech, and have a sense of humour that is either childish or unusual.

Asperger's syndrome is distinguished by a pattern of behavioural issues rather than one single symptom. Gillberg produced a set of criteria to diagnose Asperger's syndrome, by identifying traits in a person's behaviour. The syndrome can be found in different intensities from someone having Asperger's tendencies to another who clearly displays all the symptoms. Adults with Asperger's syndrome generally learn strategies to help them cope with life and these people may be regarded by their peers as somewhere from 'sometimes a bit odd' through to 'definitely very strange'.

When Gillberg's criteria are applied to Edmund Denison, although some criteria cannot easily be assessed from known historical information, the score is sufficiently high to conclude that it is highly probable that Edmund either had Asperger's syndrome or at the least, had strong Asperger's tendencies.

On the positive side, people with Asperger's syndrome have often pronounced strengths such as loyalty, honesty, reliability, dedication, and determination. Additionally, they may have particular gifts, such as artistic, mathematical, scientific, and musical abilities. Edmund certainly ticked all the boxes in this area.

It would easy to brand Asperger's syndrome as a mental disorder, but its characteristics have guided famous people to be outstanding in their particular spheres. Albert Einstein, Leonardo da Vinci, Isaac Newton, Henry Ford, and William Shakespeare have all been claimed to have had Asperger's syndrome. It seems highly probable that we can thank the peculiarities of Asperger's syndrome to have been the driving force that made Edmund the tower of the man that he was. So to relativity, the Mona Lisa, gravitation, Model T cars, and *Hamlet*, we can now add the Great Clock at Westminster as another offspring of Asperger's syndrome.

PICTURE POSTCARDS FROM 1903

The production of picture postcards experienced a boom from 1903 when the divided back was introduced: the message and address written on

The Great Westminster Clock ("Big Ben")

The Chimes of "Big Ben" are set to the following beautiful lines:—

"All through this hour, Lord be my guide, And by Thy Power, no foot shall slide."

MADE BY

E. DENT & CO., LTD.

SOLE ADDRESSES:

61, Strand, and 4, Royal Exchange, London.

FIG 14.2 A large-size E. Dent & Co. advertising postcard, dating before 1920.

one side, and a picture printed on the other. Until the beginning of World War I, the postcard was a fast and convenient method of communication. It was used for the mundane, greetings, and advertising. E. Dent & Co. introduced a series of postcards with a picture of the Clock Tower on one side and details of the movement on the other. There was little or no space to write a message, so these cards were rarely posted. At least 18 variants are known, covering combinations of different views, various addresses, and with or without the music of the chimes. These cards were probably given away by Dent; they are quite common, so they must have been produced in great quantity. Of course, the Clock Tower was one of the most photographed studies in London and thousands of different postcards must have been produced for public consumption.

OVERHAULS

The first major event in the new century was the announcement that the Great Clock would be stopped for about three weeks for cleaning in August 1900. Cleaning and overhauls of the Great Clock also took place in 1907, 1915, 1923, and 1934.

In 1912, the dials were repaired, cleaned, repainted, and gilded; the task cost £240 and involved erecting scaffolding.

1912

The chore of winding the Great Clock eventually became an issue and Dent was commissioned to design and make a winding unit, which was installed in June 1912.

FIG **14.3** Dent's men working on the clock removing the going train barrel. The date would be 1907, when some major work was carried out. Photo by the press photographer Hodsoll. From the London Metropolitan Archives.

FIG **14.4** Another Hodsoll photo of work on the clock. Note the electrical switch that can be seen above the *Fixed Here 1859* plate. From the London Metropolitan Archives.

1913

A question was asked in the House in June 1913 about the cost of winding the Great Clock. This brought the response:

As regards the clock, the expenditure of £105 is under a contract which, I believe, covers renewals as well as the winding. When he remembers how big the clock is, and the amount of regular attendance required, I think the Hon. Member will agree that 100 guineas is not an excessive charge.

WINDING UP BIG BEN

THE WINDING UP OF BIG BEN, WHICH NOW ENGAGES TWO MEN FOR SEVERAL HOURS A WEEK, IS TO BE DONE IN FUTURE BY ELECTRICITY

FIG **14.5** An illustration from *Black & White* for November 1902. Although the caption said that the Great Clock was to be wound by electricity, this did not happen for another ten years.

WORLD WAR I 1914–18

With the fear of bombing raids, Big Ben's chimes were stopped, lest they should be a guide to raiding Zeppelins.

The day after the Armistice with Germany was signed, it was asked in the House of Commons if it could be arranged for the great clock to chime and the bell called 'Big Ben' to strike the hours as before the War. In his response, the Commissioner stated, '*The striking mechanism of the Great Clock is so complicated that it will require from two to three weeks to get it into working order again, but immediate steps were taken yesterday to*

bring such portion of the chiming mechanism into oper-
ation as was possible in the circumstances, and the bell
called 'Big Ben' started striking the hours at one o'clock
in the afternoon.' It had been hoped to get the
clock striking from 11 a.m., but for some reason
this had not been possible. Dent claimed that it
was striking at noon. Big Ben was reported to be
again striking the hours on 12 November.

Fire!

An alarming incident occurred on 9 July 1918. It
appeared that the Clock Tower was on fire, and
fire engines were summoned. Fortunately there
was no danger; men had been burning wood

1918   LONDON 11 A.M. NOVEMBER 11TH.

Photo: S. & G

A scene in Whitehall at 11 o'clock on November 11th, 1918, when it was known that at long last hostilities had ceased.

FIG 14.6 Victory: 11 a.m., 11 November 1918. This
photo from a war magazine must have been posed with
Big Ben added in the background; note how the tower
base fades into the foreground.

shavings from work in Westminster Hall and the
smoke appeared to have got into the great air
shaft that vents to the belfry.

1919: CYRIL JOHNSTON

Gillett & Co. and Gillett & Bland were succeeded
by Gillett & Johnston, who started making turret
clocks around the 1860s. The company was based
in Croydon and cast its first bells for clocks in 1877.
In 1902, when Cyril Johnson was aged 18, he
joined the company working for his father. In
1906, Gillett & Johnston cast a set of clock bells for
Elstree School, which turned out to be very poor
in tone, much to the disappointment of the
customer. Cyril had an exceptionally good musical
ear and, driven by the problem of the poor bells,
he started to study bell tuning; this brought him
to Canon Simpson's system of five-part tuning. In
the 1890s, Canon Simpson devised a system that
involved tuning the five principal notes in a bell
to give a pleasing tone. Simpson's system had been
used by Taylor's of Loughborough since 1896 and
Cyril employed it to tune the replacement clock
bells. From this beginning, Gillett & Johnston
became a noted founder of bells and, to this day,
the tone of Gillett bells from the period 1920 to
1950 are highly rated by change ringers.

Gillett & Johnston bells were installed in
churches as clock bells and complete rings for
change-ringing, and many carillons were also
built. A carillon is, technically, a set of at least 23
tuned bells used for playing tunes. In Holland
and Belgium, every major town had one or more
carillons, in England, there were hardly a dozen
instruments. Bells arranged as a carillon never
gained any real popularity, here the principal
interest being in the English system of change-
ringing. However, the United States provided a

ready market for Gillett & Johnston's carillons, as did the British colonies.

Not surprisingly, Cyril Johnston turned his attention to Big Ben, of which he said, '*Its tones are enough to drive any good bell founder into an early grave.*' An article '*Carillons in England*' appeared in *The Times* of 1 August 1919. The author was 'A Correspondent', and it is pretty certain that this was Cyril Johnston. After extolling the virtues of a carillon in Cattistock, the author regretted that there was not a heavier set of bells in the country and wrote:

Surely of all the towers in the land none would be better suited for such a set of bells than the Clock Tower of the Houses of Parliament, which is the same height as the grand tower of Malines! There is ample room at the top of the tower for a magnificent carillon. The bells would be at a satisfactory height, and would be heard well over an extensive area, proximity to the Thames being an advantage in this respect. We should then be rid of cracked Big Ben and his cacophonous voice, and in his place we should get a great bell slightly lower in pitch, a true and sound casting of correct proportions, and perfectly tuned. The carillon would encompass four octaves, containing 47 bells, would weigh over 50 tons. Such an instrument would be worthy of the nation.

Since the art of the carillon and playing music on bells was little-known in England there was not much surprise that no interest was shown in Cyril's proposal.

Cyril's widow said that her husband would have liked nothing better than to get the world's best known most popular clock bell down out of Parliament's Clock Tower, take it home to Croydon, melt it down, recast it, tune it properly, and present it to England as his own. Despite attempts to promote his ideas, Cyril had to resign himself finally to the fact that he would never be allowed to do anything to Big Ben: the bell was just too popular and people did not want a new bell, let alone a carillon.

1923

At the very close of 1923, as the old year was dying, the British Broadcasting Corporation were transmitting a programme to celebrate the arrival of the New Year. Its engineers had been busy and had rigged up a microphone and equipment on the building exactly opposite the Clock Tower, where today Portcullis House stands. A cable of about half a mile in length had been run up the Victoria Embankment to Savoy Place, where the BBC had its new studio. Just as midnight arrived, the listeners had been primed to hear some music, but instead were surprised with what might well have been the first outside broadcast...the chimes of Big Ben.

So popular were Big Ben's chimes that a permanent line was connected to the Clock Tower and a microphone installed in the belfry. The BBC had initially used a piano as a time signal; this was followed by a tune played on tubular bells by an announcer. The automatic Greenwich time signal, the 'six pips', arrived in February 1924. At 3 p.m. on Sunday, 17 February 1924, Big Ben's chime rang out on the airwaves, *Urbi et Orbi*, as it were, and has continued to do so daily up to the present time.

Reporting on the Westminster Clock's performance from May 1926 to May 1927, the Astronomer Royal wrote, 'The Westminster Clock was kept purposely 2 seconds fast as the chimes were used as a warning for the Greenwich broadcast time signal. Now that the chimes are no longer used as a warning for the Greenwich time signal it is no longer necessary to keep the clock fast.' From this report it appears that, for a short period, a somewhat bizarre situation existed, where the chimes of Big Ben were immediately followed by the six pips.

FIG 14.7 Postcard showing Big Ben with the BBC's microphone highlighted. Published by Walter Scott, Bradford.

FIG 14.8 Posted in 1932, this card uses the wireless time signal as a humorous reminder that the lover's evening had come to its end. Published by Birn Bros, London.

Initially, the microphone in the belfry was surrounded with cotton wool to reduce wind noise and covered with a rubber football bladder to keep it watertight. However, pigeons soon discovered this useful supply of nesting material, so the microphone was suspended in a special box that can be seen clearly in some postcards and photos. The postcard industry was ever-quick to employ a new theme.

In 1925, Big Ben's chimes were heard on a radio in North Borneo, a distance of some 10,000 miles from London. In December 1926, Big Ben's chimes were played at 9 p.m.: previously, the Greenwich Time Signal or 'pips' were used as a time signal at 9 p.m. During World War II, the chimes heralded the 9.00 p.m. news, and became a national institution.

1934

Extensive work was carried out on the Clock Tower in 1934, which involved cladding the whole tower in scaffolding. It had been found necessary to overhaul and clean the Great Clock

LISTENIN'!

"That there's that time signal from Big Ben!"

FIG **14.9** Undated but probably published shortly after the Big Ben chimes were first broadcasted, this postcard combines topical events, wireless, Big Ben, and humour.

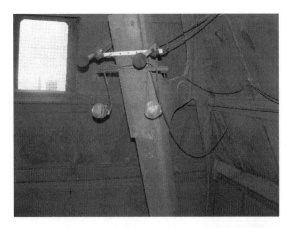

FIG **14.10** A cluster of microphones that today relay the chimes to the BBC.

and to rehang some of the chiming bells. All the bells were silent from April to July. The repairs to the clock were carried out by Messrs E. Dent & Co., the manufacturers of the clock, under their current contract, but the bell mechanism was to be overhauled by the same firm at a cost of £260, under a separate and special contract.

A.W. Hattersley

A.W. Hattersley was the Palace's resident Engineer from 1927 to 1957; he wrote a short history of the Great Clock and Bells. This was never published but several typescript versions running to about 12 pages, still exist. In his notes, he details repairs to the bells in 1934 when all the hanging bolts were renewed, the clapper inside Big Ben removed, and two of the quarter bells given a quarter of a turn so that the hammer would work on a new surface. It was possibly during this programme of work that the third bell hammer on the fourth quarter bell was removed and stowed below the bell platform.

A picture of around 1949 shows Big Ben with the clapper clearly in place but apparently supported by the wooden platform. It is believed that though the clapper was detached in 1934 it could not be removed until 1956 when the wooden platform below the bell was dismantled and replaced.

Bell Hanging

Normally clock bells are hung from a wooden beam or a headstock, either by bolts through the top of the bell, or by iron straps that hook around the canons, i.e., loops cast on the top of the bell.

FIG 14.11 A visitor to London captured with the scaffolding-clad Clock Tower in 1934.

shaped top: this top was gripped by a wrought-iron collar that must have been made in two halves. Hanging bolts from the collar passed through a plate on the top of the bell frame and were secured with nuts.

On each collar is a pair of opposing gudgeon pins, which serve as the pivot points for the hammer. Each hammer assembly comprises two wrought-iron arms that begin together and horizontal. At this end a wire rope is secured, which passes vertically downwards to the clock movement. From the end where the wire is secured, the arms divide and pass each side of the collar. Each arm is provided with a bearing so that the arm can turn about the gudgeon pins. Once past the crown of the bell, the arms sweep down and join together at the hammer head. Each hammer head is made of cast iron and can slide on the arms so that it can be positioned to strike the bell in the correct place, just on the soundbow. Oak wedges were once used to secure the hammer heads once they had been adjusted but now bolts secure the hammer head. Rubber buffers are fitted under the hammer arms, to keep the hammer head just clear of the bell, allowing the bell to sound properly.

WORLD WAR II

With the declaration of war in September 1939, a blackout was imposed in London. House, street, and car lights had to be extinguished at night to give no clue to enemy aeroplanes as to where to drop their bombs. The lighting on Big Ben's dials and the Ayrton light were all turned off.

Wellesley Tudor Pole

Wellesley Tudor Pole or WTP, as he was often known, was born in 1884. He volunteered for

The bell hammer is normally secured close to the bell in a frame; the hammer being supported in pivots and held away from the bell by a spring. In operation, the hammer tail is pulled down by a wire that goes to the clock, and then released. The hammer head falls, hits the bell and then rebounds and is held off the bell by the check spring.

The manner in which the Westminster bells were hung and the way the hammers operated was very unusual. Each bell has a flat mushroom-

191

military service in 1916, was posted to Egypt and then sent to the front in Palestine. During the evening before a battle a colleague shared his vision for a time of silence to be held each day. The soldier was killed in the battle and Pole was badly wounded. Upon recovery, he became an intelligence officer and after the war, was awarded the OBE and started his own business.

The Silent Minute

During the dark years of World War II, Pole was mindful of his late soldier friend's vision. He started to develop the idea and came up with a plan for a *Silent Minute* when all could join together in prayer, thinking about defeating the foe. Since Big Ben takes about one minute from the start of the quarter chime to the last stroke of nine o'clock, he proposed 'The Big Ben Silent Minute'.

Pole led a campaign to establish the Silent Minute Observance. A committee was formed in June 1940, who lobbied politicians, Royalty, the BBC, and religious leaders, and wrote letters to newspapers. A leaflet *'The Spiritual Front'* was reprinted many times. The June 1941 edition outlined the objectives of the movement.

THE SPIRITUAL FRONT IN UNITY IS STRENGTH

The Big Ben Silent Minute provides men and women everywhere with an opportunity to unite in dedicating their every thought, word, and deed to the service of God and freedom. Over four million people have already enlisted in this army of the inner front, and you are invited to join them, and to urge all your friends to do likewise.

It is to ensure that union in the Observance shall be on the widest possible basis that it has been made completely non-sectarian in order to embrace all denominations, faiths, and creeds. Thus every individual can observe the Silence in whatever way he or she thinks best. There is a power available to us all which we have hardly yet begun to realize. It is so infinitely great in wisdom and strength, that no combination of material forces can prevail against it. Each one can

become a channel for this power, and it can only be used if we offer ourselves to be used by it. Conscious action on our part will bring into activity the omnipotence of God against which the darkness of evil is impotent.

Eventually, the BBC agreed to the idea and on Armistice Sunday, 10 November 1940, the chimes of Big Ben were broadcast on the radio at 9 o'clock in the evening. Howard Marshall, a famous radio commentator and director of public relations at the Ministry of Food, gave an address explaining the reason for the chimes. Eventually, the BBC billed the 9 o'clock slot in the *Radio Times* as the *Big Ben Minute*,: explaining the name as follows:

It is established that millions of listeners, whatever else in the programmes they may choose to miss, tune in their sets each night to hear the solemn sonorous strokes of Big Ben at 9 p.m. What they are really awaiting, of course, is the news of the day. Nevertheless, for one minute before the news they listen to—what? A clear, a deliberate pause in the day's broadcasting. When Big Ben came back in November 1940 to a regular place in the programmes, it was suggested that this minute's pause might be used by listeners as an opportunity for rededicating themselves to the great effort in which the country is engaged, for thoughts of those who are absent, or for other individual needs. The habit of so using it has spread widely, and the adoption by the BBC of the phrase 'Big Ben Minute', which observant listeners will notice in the programme pages this week for the first time, is a recognition of the special significance that so many listeners have now come to attach to the radiated voice of Big Ben.

The proposal went down well with those who took part, participating at whatever level they chose. In April 1941, the Observance Council received messages of support from the King, the Archbishop of Canterbury, and the Prime Minister. The movement was not aligned to any religion, though its followers were generally Christian, across all denominations. Observers were invited to say the Lord 's Prayer as the Great Clock sounded, a prayer that fitted conveniently into the minute time slot.

In his book, The Big Ben Minute, Andrew Dakers gives the reason for his publication: 'It is the purpose of this little book to spread this knowledge as widely as possible, so that more and more spiritual power may be generated and released for the use of the powers of good through the more universal observance of the Silent or Dedicated Minute.' Pole believed that the dynamic power of 'thought directed sound' would produce a powerful weapon.

Pole's beliefs about the Silent Minute are revealed in a personal letter to a lady dated 10 January 1942 in which he wrote,

Regarding the B/B 'SM' work we have reached a delicate issue—we want fuller cooperation from the Govt: BBC + press + must be careful that the idea does not get about that this is an occult or magical practice; + that its sponsors are so inclined. There is a real danger of this which is one reason why the Churches + their leaders give so little support and make my own path difficult. Therefore great care is needed when speaking about the dynamic power of thought directed sound. In Big Ben with its mantric vibration we have a very powerful weapon, as you will realize from the notes I sent you yesterday. Use this information with the utmost discretion...

In the 'notes I sent yesterday' referred to he wrote, '*The power now impregnated into these chimes is not only creative + unifying but protective + can nullify black vibrations directed towards us + our cause from Germany.*'

The term '*the power of concentrated thought*' was a recurring theme from the organizers of the Silent minute.

A whole range of literature supported the Silent Minute, the Observance published at least four different postcards, along with a display card that could be hung up in factories and offices, and a Cinderella stamp was even printed, which could be attached to envelopes or letters.

Many bodies and churches joined in the Silent Minute Observance and booklets of prayers and thoughts were printed to help those who participated.

The BBC continued to use Big Ben's complete chime to announce the 9 o'clock news on the radio's Home Service until 1960. Then, on 11 May, the BBC announced that the evening news was to be switched to 10 o'clock from 19 September; citing competition from television.

Public reaction against the move was unexpected and Archbishop Fisher called it '*The ending of an epoch*'. The Council of the Big Ben Silent Minute objected, as they regarded the removal of the Silent Minute slot as a national sacrilege. The BBC Governors stuck to their decision to move the time of the news but invited comments on how Big Ben should be treated. Their final decision was that the famous chimes with the last quarter and the nine full strokes of Big Ben would be replaced with the last quarter and only the first stroke of the hour of ten, the following strokes being faded out. Eventually the chimes at 10 o'clock were replaced by the pips and today Big Ben announces Radio 4's 6 p.m. news, but the announcer cuts straight in after the first stroke, and the second is rapidly faded out. The chimes are also heard on the BBC World Service at midnight.

Wartime Damage

Anti-aircraft guns fired countless thousands of rounds into the London night skies during bombing raids. Not surprisingly, a couple of shells hit the clock tower but the damage caused was minimal.

On the night of the 10–11 May 1941, incendiary bombs were dropped that caused fires in the House of Commons and the roof of Westminster Hall. As in the fire of 1834, there was a difficult choice to be made; the fire service said

The Call

Almighty God—To Thee we call :-
*Send forth Thy love and wisdom
to lead us out of tribulation
into the ways of peace*

Help and inspire us
in the building of a better world.

Enable me to do my part
in humility and selflessness.

Guide and bless those we love
wherever they may be

Thanks be to God.   Amen.

———————————

This Call to God is made by many
immediately before the Big Ben Minute
at nine o'clock each evening.

It is suggested that this call in
conjunction with the Lord's Prayer should
be the keynote for the Silent Minute
during these momentous times.

———————————

*You may care to pin this card
near your Wireless Set.*

Copies of this card obtainable from: Big Ben
19 Bell Moor, London, N.W.3.
2/3 for 25; 3/9 for 50; 5/6 for 100 post free.

OBSERVE THE
SILENT MINUTE
WHEN BIG BEN
STRIKES NINE
EACH EVENING

FIG 14.12 A postcard inviting participation in the
Silent Minute.

FIG 14.13 This Big Ben Silent Minute stamp is known
as a *Cinderella*. It has no postal value and would have
been used to promote the Silent Minute along with
normal stamps on an envelope or other
correspondence.

that it would be impossible to save both, so it was decided to concentrate on saving the ancient Westminster Hall. The House of Commons Chamber was completely destroyed. A small bomb struck the Clock Tower and broke all the glass on its south face, but the clock and bells were undamaged and the chimes carried on being broadcast as normal.

1944

Correspondence in *The Times* for May 1944 proposed that Big Ben be recast as a war memorial. The Ministry of Works responded that,

'The matter had been thoroughly gone into some years ago and it was decided not to alter the familiar tone of the bell.' The exchange was reported in the *Horological Journal* and the leader stated, 'Actually the cracked old bell has a raucous cacophony which is a very bad example of the fine old English art of bell founding. It was never meant to sound as it does, and would not, had it not cracked, soon after being hung.' These words seem to be very much the sort of thing that Cyril Johnston of Gillett & Johnston would have said. No doubt Robbie Robinson, the principal reporter of the *Horological Journal*, knew from whom he could get a suitable quote.

FIG 14.14 A pretty little colour booklet with thoughts by 'Allan Junior', printed by Valentine; it cost just 6 d. There is a foreword by the Silent Minute Observance Council.

FIG 14.15 The bell Big Ben is the centre of this montage produced by the artist, Frank Salisbury.

FOR WHOM THE GREAT BELL TOLLS

The practice arose of tolling Big Ben at the funerals of monarchs but Queen Victoria missed this national tribute, since her lying-in-state and funeral were held in Windsor.

Victoria's son, Edward VII, was the first to be honoured, in 1910. Big Ben was tolled once every 15 seconds as the King's coffin processed from Buckingham Palace to lie in state at Westminster Hall and again when it left Westminster. It was then sounded once every minute, until it had been rung 69 times, once for each year of the age of the passing King. Big Ben was sounded 70 times for George V's funeral in January 1936 starting at 9.30 a.m. The Great Clock was then silent until midday, when the normal striking was resumed.

King George VI passed away in his sleep at the age of 56 on 6 February 1952, after 16 years on the throne. At 9.30 a.m., the funeral cortege started its journey to Windsor, proceeding to Paddington Station, and again accompanied by the tolling of Big Ben, which was tolled 56 times, at intervals of one minute, once for each year of the King's life.

BIG BEN AND CHIMES
THE CLOCK TOWER · THE PALACE OF WESTMINSTER

FIG **14.16** Frank Salisbury also produced this painting of the Westminster bells.

With best wishes to Dora & Elaine for Christmas & the New Year & many thanks for dear little calendars. Hope you will like the enclosed – Love from M. S. Walton

OBSERVE THE
SILENT MINUTE
WHEN BIG BEN
STRIKES NINE
EACH EVENING

Copies of this card obtainable from: Big Ben
19, Bell Moor, London, N.W.3.
2/3 for 25 ; 4/- for 50 ; 6/6 for 100.

FIG **14.17** Big Ben Silent Minute postcards were available from the Silent Minute Observance Council, at 1 s. 9 d. for 25. The price later rose in stages to 2 s. 6 d. for 25.

To toll Big Ben, first the quarter bell hammers would have to be disabled. Next the clock hour train would have to be released manually so that one blow would be sounded. In most turret clocks, this is a simple task achieved by removing the count wheel and lifting the locking lever to release the striking train. On the Great Clock, there are two problems; the hour hammer is left on the lift, and there is no arbor that does one turn for one blow of the bell. Most probably, the tolling was achieved by completely removing the striking locking lever and replacing it with a short one that did not engage with the count wheel. The locking lever would be held up by hand until one blow was struck; it would then be released, locking the train. A minute later the process would be repeated.

FOR WHOM THE GREAT BELL DIDN'T TOLL

A State funeral was given for the statesman and wartime Prime Minister, Sir Winston Churchill. This national event took place on Saturday, 30

'The War Illustrated,' May 30th, 1941   Registered at the G.P.O. as a Newspaper
Vol. 4 ★ ON A NIGHT IN MAY THE NAZIS CAME TO LONDON! ★ No. 91

Though Scarred 'Big Ben' Still Carries On

FIG 14.18 Big Ben on the morning after the bombing.

January 1965; the funeral was held at St Paul's cathedral. As the funeral procession left Westminster Hall and made its way to St Paul's Sir Winston was given a 90-gun salute, one shot, for each year of his life. As a special mark of respect Big Ben was silenced from 9.45 a.m., when the funeral started, and stayed silent until midnight.

1956

It was decided to carry out extensive works to the Clock Tower; this work started in 1955. The bells had been repaired in 1934, at a cost of £300, and the cast-iron roof had not been painted since 1878. It was the first time that major work was carried out on the stonework, which had suffered wartime bomb damage. One task was to deal with corrosion of the bell frame and wear to the hammer assembly, and it was also necessary to make it easier to adjust the bell hammers correctly. All the bells were lowered and the ironwork was sprayed with zinc. New collars, gudgeon pins, and bearings were provided, along with hanging bolts and top plates. The lower hammer arms were cut off and new ends were bolted on; new hammer heads were made and fitted with bolts. A modified shoe arrangement was made for the rubber buffers, allowing them to be adjusted accurately. The worn bell surfaces were dressed flat and the whole assembled and adjusted. Asbestos packing was used in the collar boxes to replace the old India rubber. Bronze was used for the bearings and bronze eyes were fitted to the wire lines and rods that connect the hammers to the clock movement. Dent was the prime contractor for the project and they subcontracted the bell work to the Whitechapel Bell Foundry.

During this restoration, the bell frame that was badly corroded was cleaned, painted, repaired, and fitted with new flitch plates. Platforms to give access round the bell heads were installed, making lubrication of the hammers an easier task. In 1862, a wooden platform had been installed under Big Ben as a safety measure should the bell break. The platform was found to be rotten, so this was removed and replaced with a new frame of iron and timber.

Because of this maintenance, Big Ben was silent from 2 July 1956 and was not heard again until Sunday, 23 December of the same year.

Extensive repairs to the clock tower and roof were undertaken; the whole tower was covered in scaffolding to enable the work to be carried out. Stonework was repaired and replaced using Clipsham stone, decorative iron work had to be recast and gilded. Since the roof of the lantern is made from cast-iron plates, a system of cleaning with needle guns was used. Once it had been cleaned down to the bare metal, the iron was treated with rust inhibitor and three coats of rich metallic lead paint, and then finished with a coat of grey micacious iron ore paint. Finally, the work was completed and the scaffolding struck during March and April 1957. The bill for the Clock Tower repairs amounted to £56,000.

Following bomb damage in 1942, the south dial had had to be reglazed. In July 1956, extensive work on the Palace included rehanging the bells, an overhaul of the clock, and reglazing all the dials.

1959

It was decided to celebrate the Great Clock's centenary, commencing in June 1959. A stone plaque was unveiled on the Clock Tower base; this can be seen from Bridge Street. A wooden copy was made for display in the clock room.

An exhibition was organized, comprising a display of a full-size drawing of the clock, portraits of Sir Benjamin Hall, Barry, Dent, and Denison, a small turret clock by Dent, and a model of the gravity escapement. This model was made by Messrs Cohen and West of the British Horological Institute and is now in the BHI museum collection at Upton Hall. An accurate model of the Clock Tower made by James Mabey, who was a stone carver to Barry, wooden patterns used to make iron castings, and other ephemera connected with Big Ben and the Clock Tower were also shown. The exhibition was held in the Westminster Jewel Tower, which is within a few hundred yards of Big Ben and is

FIG 14.19 Model of Denison's gravity escapement exhibited at the centenary exhibition.

E. DENT & CO. REDUCES ITS ACTIVITIES

Thomas Buckney married the sister of Frederick Dent in 1838. After the deaths of Frederick and Elizabeth Dent, Thomas inherited a half share in the company of E. Dent & Co. The Buckney family continued in Dents until 1968, when Patrick Buckney, the managing director, retired.

The January edition of the *Horological Journal* for 1971 records a gift to the Science Museum of a Dent three-train turret clock; as the company was in the process of reducing its size, they gave up servicing turret clocks and closed their factory. The Dent company name was bought by Toye & Company in 1974, and Dent stayed in its Pall Mall address until 1977; a key part of its business was making reproduction clocks for collectors.

E. Dent & Co. Withdraws from Maintaining the Clock

In February 1971, a question was asked in the House of Commons about the maintenance of Big Ben, since there had been some problems. The reply was that after more than a century of service, the company (i.e., E. Dent & Co.) that wound and serviced the Great Clock had decided that they could no longer maintain the clock. A new company was being sought to do the job. On 4 May, an article appeared in *The Times*, announcing that Thwaites and Reed were the company to take on the winding and maintenance of the Great Clock. They described how John Vernon, Thwaites's Turrets and Specials Manager, climbed the Clock Tower and wound the clock for the first time. Thwaites and Reed were established in 1740 and operated in

itself one of the last remaining fragments of the mediaeval Palace of Westminster. The exhibition continued throughout the summer until the end of September, during which time the Clock Tower was floodlit. Mr Speaker made an address in New Palace Yard on the occasion of the Great Clock's centenary.

1960s

There was a suggestion that Big Ben be silenced at night for the benefit of patients in St Thomas's hospital, which is just across the river on the south bank. However, a survey showed more patients seemed to be comforted by the chimes than were annoyed by them.

Clerkenwell. In the 1970s, they moved out of London to a factory in Hastings.

1976

In August 1976, a catastrophic failure occurred that severely damaged the quarter train. Repairs took over nine months to complete. Chapter 15 covers this topic in detail.

1982

In the 1980s, an extensive programme of stone cleaning and restoration was carried out at the Palace of Westminster.

1984: COLOUR OF THE DIALS

Extensive restoration work was carried out on the Clock Tower in 1984. In reply to a question asking if the original colours were to be used on the dials, the Secretary of State for the Environment said that,

The colour scheme for the dials would be retained. He continued to state that there were no original drawings as to the colour of the dials but the main cast-iron divisions of the dial and letters were described as blue, in a book bearing Sir Charles Barry's name, but published 5 years after his death. Paint samples indicated that there have been a number of differing colour schemes; a sample from the lower part of the clock dial appeared to show five schemes of which the more recent was black. The existing black and gold scheme was at least 50 years old, more than a third of the life of the building. These are the colours with which the world has become familiar. For this reason and because of conflicting and inconclusive evidence of previous schemes, my Right Hon. Friend decided that the existing colours should be retained.

1990: A NEW HAMMER FOR BEN

In 1990, cracks were noticed in the frame arms of Big Ben's hammer. The hammer comprises a head of cast-iron, about 3 cwt; this is held by two long wrought-iron frames that sweep down from their pivots close to the crown of the bell to the head. In the other direction, away from the pivots, the frames extend more or less horizontally, meeting where the steel line from the clock connects. Each of the two halves of the hammer frame is a single forging. The Whitechapel Bell Foundry was charged with the repairs, and two new forgings were made to do this. When the bells were first installed in 1858, forges that could deal with large pieces were quite common. With changes in modern manufacturing techniques, forging large items is almost a thing of the past. However, Whitechapel were able to get the work done and the new hammer arms were duly installed. In response to a question asked in the House of Commons, it was revealed that the cost of the forgings was £8,000.

1993

The mechanism of the Great Clock is subject to regular expert inspections and routine maintenance. Included in the inspection programme for 1993 was the replacement of the quarter-striking great wheel. No doubt, non-destructive tests carried out by Harwell had identified cracks in this wheel, which was made in 1976 following the major disaster. The work took about eight weeks to complete and during that period the quarter hour chimes were silent but the clock itself and the hour strike were not affected.

THE LEANING TOWER OF WESTMINSTER

The Clock Tower's foundation is a 12 ft thick raft of concrete. This stands on a stratum of terrace gravels below the top layer of alluvium. Below the gravel is London clay. When the tower was built, a plumb bob was dropped down the Clock Tower stairwell by the builders; the brass plate at the bottom that marked the centre point can still be seen. In 1960, the tower was 4 inches out of true, not bad for a 300 ft high building. Early in 1961, the lean was measured (possibly more accurately) at 14 inches and it was decided to monitor the tower regularly. In 1968, a much more accurate measurement was made; this showed a lean of 9½ inches towards the north-west. In 1972, the lean was unchanged.

An underground car park was constructed in New Palace Yard in 1972. In 1971, routine survey work found cracks in two piers that support the Clock Tower. These were duly strengthened; the cracks were not caused by the excavations for the new car park.

In the 1970s, building started on a new Underground train route to relieve pressure on the Bakerloo line. It was initially called the Fleet line but the name was changed to the Jubilee line in 1977, to honour the Queen's Silver Jubilee. Part of the route ran from Green Park to Westminster and on to Waterloo. At Westminster, there would be two new tunnels that ran underneath Bridge Street and a deep box under Portcullis House to take to the escalators that provide access to the platforms. Civil engineers know that excavations in soft ground cause changes in nearby buildings, both subsidence and sideways movement. With the new tunnels being about 100 feet from the clock tower and 120 feet deep,

a definite hazard was identified. The subsoil in which the tunnels were to be cut was London clay, above which was gravel, then alluvial material. Computer modelling predicted that the likely movement of the Clock Tower would be unacceptable, so a technique of correcting the tower's movement by compensation grouting was used.

The solution was fairly complex. The first step was to employ a construction method that created the minimum soil disturbance. Tower movement had to be accurately known, so surveying sensors were fitted to each corner of the tower, and an electronic plumb bob installed inside the clock tower. These were connected to a computer, so that the movement of the tower would be accurately known.

A vertical shaft was sunk in Bridge Street approximately where it is joined by the Victoria Embankment. The shaft went down to a level just into the London clay; from this shaft about 15 horizontal tubes were bored that fanned out below the clock tower foundations and under Bridge Street. Using controllable rubber sleeves and packers, grouting (a sand and cement mortar) could be pumped to any required position along any of the tubes. The mortar grout caused a blister in the subsoil that not only prevented further movement but pushed the tower back into an upright position. It was decided that corrections had to be applied if the tower moved more than 15 mm (about ½ inch) out of upright. When the sensors detected this, appropriate injections of grout were made and the lean was stabilized. Once all the grouting corrections were completed, the lean was within 15 mm. Ground movement continued, as the London clay adapted to new pressures and water levels and settled into a long-term equilibrium. Two years after the last grouting, the lean stabilized at

35 mm (about 1½ inches), which was within acceptable limits. Regular monitoring of the clock tower lean continues today.

STOPPAGES

Of course, a mechanical clock will have problems from time to time. In general, the issues have been with the bells, where wires and their associated shackles have broken or come undone. Rubber buffers that keep the hammer head just off the bells have frozen in cold weather and become so hard that the tones of the bells have been affected.

Regarding the clock, probably the chief enemy is the weather; particularly, snow. Wet snow accumulating on the hands is heavy, and this problem becomes worse when, the harder it freezes. Heating inside the dial rooms helps the situation but it's a large area and any amount of heat may still not defrost the hands sufficiently for them to work.

The following is a list of stoppages and problems up to 2000. Not every event is recorded but the list is given as a sample.

1906		Clock stopped; cleaners had left a plank against a wall, jamming the motionwork's counterbalance
1907	May	Clock overhauled
1912		Clock stopped
1915	March to April	Clock overhauled
1923		Clock stopped for overhaul
1934	March	Clock stopped for overhaul: St Paul's cathedral clock used to broadcast chimes on radio
1936	September	The Clock was stopped, allegedly by a painter's ladder leaning against the hour tube whilst the dial chamber was being whitewashed: this is possible, but unlikely; it was more likely that the ladder was in the Clock Room and the counterbalance of the minute hand fouled on the ladder
1937	January	A shackle on the wire to the hour bell broke; presumably the wire fell into the going train and stopped the clock
1941		A workman's hammer was wedged in the works
1944		Broken suspension spring
1949	August	A flock of starlings settled on the hands; their combined weight made the clock run five minutes slow: the time must have been about a quarter to the hour, with the birds sitting on the minute hands
1952	February	Chimes were disabled whilst new rubber buffers were fitted to the bell hammers
1954	September	It was announced that extensive restoration work was to be carried out on the Clock Tower over the next two years
1956	July	Chimes disabled for overhauls

1956	November	Overhauls of clock and bells completed
1956	December	Chimes reinstated
1957	May	New lights fitted to the dials
1957	November	Hour striking failed because a wire had pulled out of a shackle
1961	October	The Clock failed to strike the hour because of a fault in the winding mechanism
1962	January	The Clock was ten minutes slow because of snow accumulating on the north dial
1966	February	The Clock failed to strike the hour; this was caused by a loose shackle pin
1966	June	The Clock stopped striking the hours because of a defect in the winding mechanism: a motor had not cut out and the clock was overwound, breaking half a tooth in the hour-striking train
1966	October	A note was missed out on the quarter, a rod had become disconnected
1966	November	The Clock stopped for 22 minutes, as a mechanic had inadvertently 'flicked over a ratchet on the winding mechanism'
1972	March	The Clock was running 15 seconds slow after the switch to summer time: the reason given was that a new spring fitted was 'bedding in'; this must have been a new suspension spring for the pendulum
1976	January	The chimes were silenced so that a room in the Clock Tower could be redecorated
1976	January	The Clock stopped: the engineers were unable to find any specific fault
1976	August	The quarter train was almost completely destroyed
1977	May	The Great Clock's quarter striking was restarted
1977	May	The Clock stopped because of a lack of oil on a bearing
1978	April	The Clock Tower was to be restored; it was in scaffolding for 18 months
1983	March	Scaffolding erected on Clock Tower
1985	May	Top of Clock Tower unveiled
1985	December	All scaffolding dismantled
1986	March	Cold weather freezes the rubber buffers on the clock hammers and silence the chimes
1986	March	Hammer buffers froze
1987	June	Excavations for new underground car park were thought to cause Clock Tower to lean
1987	October	Hour striking silenced after non-destructive testing when the Clock was put back from summer time: cracks found in bracket holding hour fly blades
1990	March to August	Fatigue cracks found in Big Ben's hammer
1990	August	Clock stopped: no problem found
1994	January	Clock stopped: no problem found

On 12–13 January 1987 the cold weather caused a problem with the chimes of the Great Clock. The normally flexible rubber pads that hold the bell hammers just clear of the bells became hardened in the extremely low temperatures in the belfry. This affected the hour striking bell, Big Ben, and also one of the quarter-chiming bells that failed to sound on a number of occasions, the hardened pads preventing the hammers making contact with the bells. A method was found of warming the pads sufficiently to restore their flexibility; this was believed to be the application of a hair dryer.

CHAPTER FIFTEEN
THE DISASTER OF 1976

THE DISASTER

At 3.45 a.m. on Thursday, 5 August 1976, a policeman on duty near the Clock Tower reported hearing a muffled boom. Investigation did not reveal anything but it was soon obvious that the Great Clock had stopped. The policeman reported the problem, and the noise he heard, to the control room, where the engineer in charge thought a possible explanation was that one of the weight lines had broken. He informed the BBC that the chimes were out of order and climbed the Clock Tower to assess the problem. On entering the clock room, he saw a scene of total devastation with pieces of clock everywhere. The Bomb squad was called and Thwaites & Reed informed.

Vic Adams was Thwaites & Reed's clock-maker on call and he was soon driving across London in his van. Vic later described that day.

I got a phone call at 4.20 a.m. and was told that there were big problems with the 'Big One'. I called Albert Fairey, who used to wind the clock and told him to meet me at the bottom of the tower and not to bother to bring any tools. It was a lovely August morning as I drove into the Palace. We met Mr Butler and some policemen, one of whom had gold braid round his cap. We all went up the tower and the policemen went into the clock room but Albert and I held back. It was not that long ago that there had been a terrorist bomb at Westminster. Butler said to come in but we said it might be a bomb, and thought it sensible that we should wait. He agreed. Shortly a man in civvies arrived carrying a little wooden box; he was a Major in the bomb squad. He

asked everyone to go somewhere else for bit, so we all climbed up to the belfry. The bomb man called us back and said it was not a bomb, the windows had not been blown out and there was no trace of explosives: it was purely a mechanical failure, then he left.

We had to leave everything as it was until lots of photos had been taken, then the press came in. My colleague told them it was a bomb so the Palace had to contact the news-papers to make sure that story did not get out. After the press photographers had left we noticed that all the pennies used for regulating the pendulum were missing, stolen as souvenirs.

The whole scene did indeed look as though a bomb had gone off. From the going train to the end of the quarter train, the clock frame was mostly demolished and no wheel work remained inside. The front lower part of the frame was cracked and sagging. There were bits of broken cast iron everywhere; indeed heavy fragments had been thrown through the thick timber ceiling and were found in the link room. Every wheel was broken; the quarter barrel had come right out of the frame and had lodged itself in a wooden seat cutting a deep groove. Underneath it were some parts of the quarter great wheel. A few arbors stood against the frame, as though someone had tried to tidy up, and a broken winding wheel lay at the end of the frame.

In comparison, the going train was relatively unharmed but the going barrel was damaged and the escapement arms were badly twisted. Flying pieces had caused incidental damage to

FIG 15.1 The Great Clock on the morning of 5 August 1976: photo courtesy National Physical Laboratory.

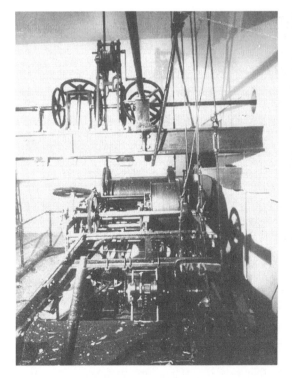

FIG 15.2 The quarter-striking train end on the morning of 5 August 1976: photo courtesy of John Wilding.

FIG 15.3 John Vernon and Les Butler examine the damaged barrel: photo courtesy of John Wilding.

the striking train. Reports of the damage were published in the *Horological Journal* and *Antiquarian Horology*.

THE DAMAGED CLOCK RESTARTED

John Darwin was the resident engineer at the Houses of Parliament at the time and was

FIG 15.4 A pile of damaged parts: photo courtesy of John Wilding.

responsible for the Great Clock. Very quickly, he arranged for the Great Clock to be restarted and to display the time on the dials to the public. It was very fortunate that during a previous repair the old escapement had been replaced with a new one and the original escapement repaired and kept as a spare. Screw jacks were used to support the broken frame as a temporary but effective fix and the spare escapement refitted to the clock. By nightfall, the Great Clock was again telling the time but without striking the hours or quarters. Such prompt action was a great tribute to the staff of the Palace and Thwaites.

Early on, the cause of the failure had been thought to be the breaking of the quarter fly arbor, so the fly arbor on the striking train was replaced as an emergency measure, probably between 16 and 27 August, when the clock had to be stopped for emergency work.

John Darwin wrote a book about the Great Clock, *The Triumphs of Big Ben*, published in 1986. This book contains first-hand information about the disaster of 1976. Much of the detail about the clock restoration in this chapter was précised from John's book.

REPAIRS

It was necessary to decide whether the clock could be repaired or should be replaced with something modern and electronic. Robert Cooke, MP, who had always had a keen interest

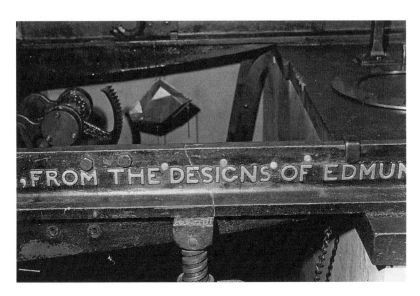

FIG 15.5 A jack supports the broken frame allowing the going train to run: photo courtesy of John Wilding.

in the historical aspects of the Palace chaired a committee to decide the clock's future. Fortunately, it was decided that the clock would be repaired, though this meant a total rebuild of the quarter train. The target date for completion was May 1977, when Her Majesty the Queen was to visit Parliament. Robert Cooke's committee was acutely aware, not only of the unique nature of the Great Clock, but of the fact that Big Ben represented the country and simply had to be repaired.

As an interim measure, on 17 August, the BBC rigged up microphones in the belfry at St Paul's cathedral: the striking of Great Tom was to be broadcast to replace the sound of Big Ben, whilst repairs were being carried out.

John Darwin was a person who got on and made things happen. Very soon, Thwaites and Reed were involved in planning the restoration of the Great Clock. They were fortunate to have as a technical director John Vernon, who had an exceptional mechanical talent and plenty of drive. The project was not simple; no

drawings of the clock existed, so the first thing that had to be done was to produce a complete set. New wooden patterns had then to be made, from which iron castings would be produced. Quite a lot of the original patterns for the Great Clock had survived and were kept in store. Unfortunately, these had suffered with time; the wood had expanded and contracted with weather changes, the glue had given way and the patterns had been damaged as they were moved around. The only solution was to make new patterns.

Once the patterns had been made, castings were obtained, then these had to be machined. Originally, the wheels in the clock were made of cast iron, the teeth were cast in and needed minimal finishing. It was decided that the great wheel and winding wheels should be made of steel, which was first to be cast then have the teeth cut in. The tooth form on the Great Clock was cycloidal, but it was decided to use the modern involute form instead. Parts were tested before installation; one such test showed a defect

FIG 15.6 Wooden patterns and a core box for preparing castings of replacement parts.

in the first great wheel casting that caused it to be rejected.

Thwaites and Reed's factory in Hastings was a hive of activity, with John Vernon masterminding the rebuild. Long hours were worked to meet the deadline of May. Forgings and castings were radiographed before being released for use; this was very useful, since the great wheel had to be recast twice before a satisfactory casting was made.

In due course, the Great Clock's quarter train was rebuilt and assembled in Thwaites and Reed's factory. The assembly was then dismantled and installed in the Clock Tower at the beginning of April 1977. One particular problem was the marrying of the new frame to the old and the provision of an appropriate support. The temporary jacks were removed and a strong but discreet steel frame installed below the clock to take the huge weight of the clock at the point of the new joint.

Finally, the bells were tested on the Sunday before the Royal visit; Sunday was chosen, since the ringing of the clock bells would not be too obvious and would be masked by the ringing of the Abbey bells. Some problems were identified and located, such as tight joints in the link room. The bell cranks were freed and lubricated. During this process of checking the clock and whilst climbing the stairs, John Darwin had a heart attack and was rushed off to hospital. Fortunately, he survived.

THE RESTORED CLOCK RESTARTED

On Wednesday, 4 May 1977, Her Majesty Queen Elizabeth visited the Houses of Parliament to receive a Loyal Address from the two Houses on the occasion of her Silver Jubilee. On her arrival at midday the Great Clock again started chiming the quarters and sounding the hours. The restoration was finished. The quarters had been out of action for eight months. Listening to the clock from his bed in Westminster Hospital, John Darwin heard the sound of the restored chimes; no doubt it was the best medicine he could have.

EXHIBITION

Once the restoration was complete, a small exhibition about the clock and the recent repair work was put together. This was exhibited from the end of July in the Upper Waiting Hall and then it moved to Westminster Hall, where it was on view to the public until the middle of October. Photos of the damaged clock were shown, as well as various broken parts.

THWAITES AND REED AND THE NATIONAL ENTERPRISE BOARD

In 1976, apart from maintaining Big Ben, Thwaites and Reed were servicing over 300 turret clocks, installing automatic winders on turret clocks, and making reproductions of antique clocks. These reproduction clocks were limited editions for collectors, many of whom hoped that their purchases would increase in value. Unfortunately, there were rumours in the horological world that Thwaites and Reed were having financial problems, which, if not solved, could have compromised the restoration programme of the Great Clock.

Thwaites and Reed wanted a cost-plus contract to do the restoration; not an unreasonable request, considering the complexity and urgency of the task. The Contracts Directorate were most unhappy with the situation but, in his book, John Darwin tells that he secured an agreement with John Vernon that in the event of Thwaites and Reed folding, he would continue to work on the clock. Much credit is due to John Darwin for having both the courage and vision to order that the repairs on the clock proceed. Progress was not always smooth, Thwaites and Reed's managing director underestimated the size of the task and, in one instance, Westminster had to buy the company a lathe of sufficient size to do the large turning required.

Financial problems continued at Thwaites and Reed: these were solved when the National Enterprise Board (NEB) essentially purchased the complete company. The National Enterprise Board was set up in 1975 to implement Labour Government policy of bringing more companies into public ownership; it was a most convenient instrument to solve the financial problems at Thwaites and Reed. On 25 March 1977, it was announced that the National Enterprise Board was to acquire 240,000 £1 shares in Thwaites and Reed, equivalent to 90% of the company's equity. Later exchanges in the House of Commons, recorded in *Hansard*, referred to Thwaites and Reed as being bankrupt. After around 130 years of being managed by one family, Thwaites and Reed passed from private into public ownership.

Thwaites and Reed carried on trading after the Great Clock was restarted in May 1977, but by December 1978, *The Times* carried an article with the caption '*NEB winds up its tick facility for T&R.*' The National Enterprise Board announced that it was selling its share in Thwaites and Reed to F.W. Elliott for £78,312. Their initial investment, plus more ploughed in to cover Thwaites

and Reed's trading loss in 1977, meant that the National Enterprise Board had lost almost £450,000 in the public ownership venture, which had lasted only 21 months.

The Opposition asked questions in the House of Commons along with a demand for a public enquiry into the National Enterprise Board's activity. Somehow, all those involved managed to miss the key fact that had Thwaites and Reed been allowed to become bankrupt during the time they were working on the Great Clock restoration, the repair work would have been halted and it would have taken a very long time before another company could pick up the project. The National Enterprise Board's action to bail out what was a small and failing clock company was justifiably open to criticism; but the fact that the Board had rescued Big Ben's restoration was unfortunately completely overlooked.

FAILURE INVESTIGATION

The National Physical Laboratory (NPL) at Teddington was consulted and contracted to investigate the cause of the failure of the Great Clock. Once the debris had been photographed, the investigation could begin. Parts were removed and examined at Teddington: key to the investigation was the vertical 12-ft long fly arbor.

The investigation team was led by Norman Owen, the Head of Engineering Services, and included Professor Cedric Turner, an internationally recognized authority on fracture toughness, who happened to be on secondment to the National Physical Laboratory from Imperial College at the time of the failure; Dr Phil Irving, an expert in fatigue; and Malcolm Loveday, who was already familiar with turret clocks, since he wound and maintained the clock at his local church and also had an interest in antiquarian horology. Like most horologists with a knowledge of turret clocks, he was able to advise the team of the probable sequence of events and, therefore, which critical components should be sought amongst the debris; in addition, before visiting the clock tower it was possible to expedite the failure investigation by providing the team with a drawing of the clock from Grimthorpe's book, *A Rudimentary Treatise on Clocks, Watches, and Bells.*

The National Physical Laboratory delivered preliminary conclusions in August 1976, followed by a letter in September, summarizing the investigation, which was almost complete by that time. The cause of the failure was given as a fatigue fracture in the quarter fly arbor. The letter concluded by speaking about the Great Clock, '*That a case probably exists for complete replacement. In view of the intrinsic and sentimental value of the mechanism, it could be preserved as an exhibit and replaced by a modern mechanism of higher reliability and safety.*' Fortunately, they had missed the boat because the Great Clock repairs were already well under way.

In its final report to the Palace, dated December 1976 the National Physical Laboratory stated, '*The primary source of the failure was considered to be a fatigue failure of the wrought-iron tubular shaft of the fly governor that controls the speed of the striking.*' The remainder of the report detailed the various metallurgical investigations that had been carried out and calculations regarding fatigue failures.

It was found that the lower part of the fly arbor was made from a piece of wrought-iron tube about 8 ft 8 inches long, thought to be a piece of gas pipe. This tube had a bore of 1 inch and a wall thickness of around 3/16 of an inch. A bevel gear was fitted to the lower end and the top part was extended to carry the fly blades. In all, the assembly was just over 14 ft long. The iron tube had fractured about 40 inches from the lower end. This led to the quarter-striking train running free, with no speed regulation; the driving weight, of 1 ton, then rapidly descended to the ground in an uncontrolled manner, causing the catastrophic failure of the Great Clock's quarter-striking train.

The iron tube used as the fly arbor had been manufactured by taking a long strip of wrought iron, heating it to red heat, and wrapping it round a long iron mandrel to form an approximate tube. The butted ends were then hammer welded to make a secure joint. From the regularity of the hammer weld it is obvious that this was done with a mechanical hammer. A better joint would have been a scarf weld, where the ends were overlapped before being hammer welded. Whilst such a system of manufacture was suitable for a gas pipe, it turned out to be quite unsuitable for a tube in torsion.

Other faults were also found. One spoke in the lower bevel gear showed an old fracture and a similar crack had started on another spoke. The locking piece that fitted onto the wheel that had the bevel drive had a fatigue fracture where the lever changed section and there was a sharp corner. A further fatigue fracture was found in a guide that steadied the locking lever.

The arbor had broken about 40 inches from its lower end; the break showed all the characteristic signs of a fatigue fracture. When the broken arbor was cut longitudinally opposite the weld, the tube fell into two parts; the weld had completely failed. Metallurgical examination revealed that the weld was of poor quality with a lot of slag inclusions. Since the tube had a covering of black iron scale, was oily and the crack propagated along the line of the weld; visual checking would have not identified the problem. In manufacture and assembly, the end of the tube had been bored out, then forced onto a slightly tapered spigot on the lower bevel gear, and secured with a pin. This method of fixing may have sprung the weld and caused the crack to start. At some stage, the tube must have been found to be loose on the bevel wheel, since a clamp had been added to hold the tube tight. It appears that the clamp had been installed early in the life of the clock.

What had happened was that the crack had started at the bottom of the tube and slowly propagated up the tube along the line of the substandard weld. Every time the clock started to strike the quarters, a torsion load was applied to the tube; this load was then relieved when the striking finished. When the longitudinal crack had propagated about 40 inches from the bottom, it reached the point of maximum flexure of the tube. The tube was supported about halfway along its length by a steady in the form of four rollers, but since the tube was not true there had

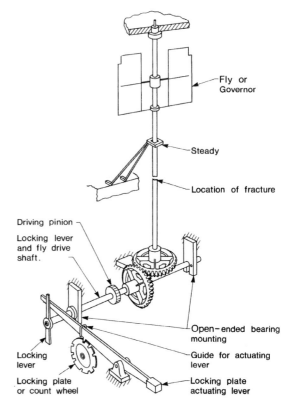

FIG 15.7 A diagram of the fly arbor and its bevel gear. Courtesy of the National Physical Laboratory.

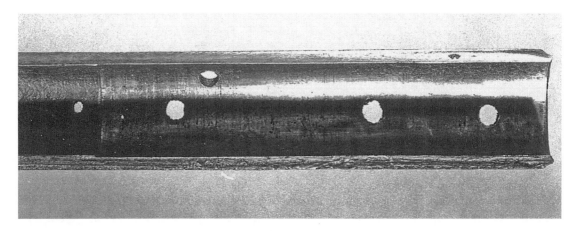

FIG 15.8 The lower end of the fly arbor: the upper edge was sawn; the lower edge is the fracture. Courtesy of the National Physical Laboratory.

FIG 15.9 The clamp that secured the cracked tube onto the bevel gear. Courtesy of the National Physical Laboratory.

to be some clearance. I remember hearing the loud rattling of the arbor inside these rollers on my first visit to the Great Clock. At the point of maximum flexure the crack divided into two cracks, which then propagated circumferentially round the tube. When the cracks had progressed far enough, the tube became too weak and broke. Calculations performed by the National Physical Laboratory showed that the tube was more than strong enough to do the job it was intended for, provided that the tube was not damaged. At the time of the investigation, the crack along the fly arbor was the longest fatigue crack recorded in a component.

Later analysis of the hour-striking fly arbor tube showed that the weld was of an even poorer quality than that on the quarter fly arbor; however, it had not cracked. Its early replacement after the disaster was a timely move.

The likely failure sequence was deduced from the damaged parts and runs as follows:

The fly arbor probably finally fractured as the clock finished striking the quarters at 3.30 a.m. The fly arbor fell to one side and the lower bevel gear became mostly disengaged with its mating bevel gear. When the quarters started at 3.45 a.m., the clock quarter train began to run away, and the teeth of the fly bevel were bent by the rapidly rotating bevel gear. Cams on the quarter great wheel struck the hammer levers at high speed, breaking cams and levers. When the train locked, the locking piece broke along the line of a fatigue fracture that had not completely cracked all the way through and the train again started. From this point, the weight would have descended and more damage caused to the quarter train that would have been running at high speed. It was later suggested by engineers who had observed failures in cranes that once the weight hit the ground, the barrel would have

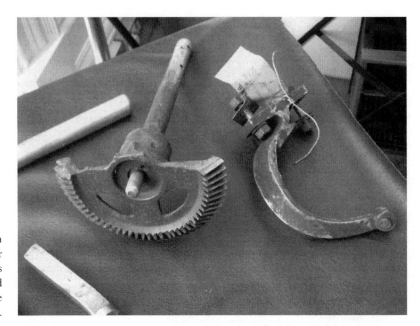

FIG **15.10** The quarter train bevel gear. The fly arbor fitted onto the spigot on this wheel. Note the damaged teeth; this was caused by the rotating mating bevel gear.

In their report, the National Physical Laboratory proposed that non-destructive testing be carried out on all the clock movement and some safety mechanism be installed to prevent a similar occurrence.

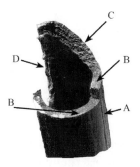

FIG **15.11** The fractured end of the fly arbor: A, hammer weld; B, tangential fatigue fracture cracks; C, helical mode fatigue crack; D, final failure fracture. Courtesy of the National Physical Laboratory.

NON-DESTRUCTIVE TESTING

Non-destructive tests began on 23 October and were carried out by the Atomic Energy Research Establishment, which is known colloquially as Harwell after its location; their team was led by R.W. Parish. Radiological techniques were used. These were similar to using an X-ray machine, but used a radioactive isotope, iridium-192, as a source of gamma radiation. Ultrasonic testing and magnetic flaw-detection tests were also used, along with fluorescent dye penetrants. Even good old-fashioned paraffin and chalk dust were used.

been rotating very fast. The remainder of the wire line on the barrel probably unravelled and was then wound up by the spinning barrel. Once the tension had been taken up, the most likely scenario was that the sudden jerk on the barrel as the line tightened caused a severe shock that broke the frame and flipped the barrel across the clock room.

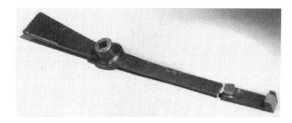

FIG 15.12 The broken locking lever. The integral leaf spring was intended to reduce the shock of locking. Courtesy of the National Physical Laboratory.

FIG 15.13 The broken end of the locking lever. Courtesy of the National Physical Laboratory.

The programme of tests revealed what was sound but the paraffin and chalk dust technique showed that there were cracks in the two bevel gear wheels in the striking train and a locking lever. For safety reasons, the resident engineer directed that the striking be suspended until the two bevel gears could be replaced. Thwaites and Reed soon produced a new set of gears; these were made of bronze, which is much more durable than cast iron. A large number of minor cracks and blowholes in various castings were revealed. The holes dated from when the clock was made, and had been filled and painted over. These were not considered problematic. Cracks were also found in the cam ring on the great

wheel of the striking train. A special reinforcing ring was made and shrunk on to give additional strength.

In all, 13 castings were identified as faulty and replaced. As part of the on-going maintenance of the Great Clock, regular non-destructive testing is still carried out by Harwell.

PREVENTION IS BETTER THAN CURE

Westminster commissioned the National Physical Laboratory to design and make a safety brake. Jim Furze and Norman Owen produced a device that was neat, compact, and effective. Classically, the basic design concept for the safety brake was sketched on the back of an envelope by Jim Furse, in discussion with Malcolm Loveday, whilst they returned on a train between Waterloo and Teddington. They had been up the clock tower to consider the practicality of fitting some form of device to the clock movement to prevent a similar failure happening in the future. Basically, the brake comprised a pinion that was driven by the winding wheel on the barrel and a pair of brake pads that spanned the winding wheel rim. This pinion drove a bevel gear that turned a cup-shaped cylinder with a curved bottom. Inside the cylinder was a ball bearing, and, as the speed of the cup increased, the ball bearing rode up inside the cup. If the speed of rotation passed a certain maximum figure then the ball bearing passed over the top of the cup and dropped down, engaging the drive to the brake calliper, which was then applied by the turning motion of the winding wheel through the pinion.

Two brake units were made and fitted, one to the quarter train and one to the hour train. No

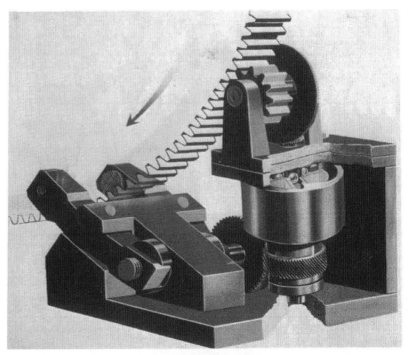

FIG 15.14 The pinion, top right, drives the speed sensor. When over-speed is detected, the drive is connected through the lower helical bevels to engage the brakes. Courtesy of the National Physical Laboratory.

The 'fail-safe' mechanism

safety brake was fitted to the going train, since, in the event of a problem, the speed of running would be significantly slowed by the load of the hands.

THE COST

A question had been asked in the House of Commons concerning the cost of repairs to the Great Clock but it was stated that these costs would not be known until a later date. The question was not asked again so the costs remain, probably fortunately, unknown.

FAILURE CONFERENCE

The Institution of Mechanical Engineers staged a one-day symposium on 11 June 1981: its subject was the 1976 disaster and the title was *Big Ben—Its Engineering Past and Future*. Papers read included a history of the Great Clock that I had written, as well as papers on the failure of the quarter chimes, the safety brake, and the non-destructive test carried out. A small booklet was produced to commemorate the event.

RESURGAM

(Latin: I shall rise again)

Thwaites and Reed, under its new ownership of Elliott in 1977, continued to trade under its old name, working on prestige turret clock installations and maintaining the Great Clock. In the 1990s, a fall in demand for domestic mechanical clocks left Elliott short of business. This led to the company being bought out and the Elliott clock manufacturing side was subsequently sold off.

Thwaites and Reed was then back to its core business of turret clocks that it had started in 1740.

A brief article in the *Guardian* of 26 March 2002 revealed that Thwaites and Reed had been unsuccessful in renegotiating its contract with the sergeants-at-arms, who were responsible for daily management of Westminster. Two Thwaites and Reed clockmakers joined the Palace staff and continued to wind and care for the Great Clock. Thwaites and Reed had had over 30 years of service caring for the Great Clock, during which time the company

FIG **15.16** John Vernon, Thwaites and Reed's technical director. Photo courtesy of John Wilding.

FIG **15.15** John Darwin, resident engineer of the Houses of Parliament. Photo courtesy of the British Horological Institute.

masterminded the major restoration of rebuilding after the disaster.

ACKNOWLEDGEMENTS

Two people that may certainly be thanked for getting the Great Clock back into action after the catastrophic disaster. The first was John Darwin, the resident engineer, who drove the project forward and made things happen against all odds. The second was John Vernon, of Thwaites and Reed, who worked with skill and dedication, organizing and running the clock restoration.

CHAPTER SIXTEEN
THE RESTORATION OF 2007

THE NEED FOR RESTORATION WORK

Extensive maintenance work was needed in 2007 because wear had been noticed in the striking train where the strike barrel was seen to be rubbing on the great wheel during winding. On the going train, the lantern pinion trundles on the escape wheel pinion were becoming deeply pitted, so that was another indicator that work was needed. The motor-assisted winding mechanism installed below the movement was showing wear on a pinion. No significant work had been carried out on the striking or going trains for many years. Of course, since the Great Clock is such an important icon of the British nation, it is not possible to take it out of service at the convenience of the repairer as one can do for a church or town hall clock.

After extensive planning by the Keeper of the Great Clock, much of the work was carried out by the Palace Clockmakers, Ian Westworth, Huw Smith, and Paul Roberson. Specialists were used for scaffolding, lifting, and heavy machining. As part of the forward planning it was decided to keep the hands telling the time so it was necessary to specify a drive unit that would neatly fit next to the bevel gear cluster. It had to be very powerful but also easily engaged and disengaged as needed. A special synchronous unit was commissioned, built, tested at full load,

and installed in the tower the week before the major work started.

RESTORATION STARTS

On the morning of Saturday, 11 August 2007, a goodly collection of the media gathered on the roof of the House of Commons to witness the start of an extensive programme of maintenance on the Great Clock. With the 150th anniversary of the clock just over a year away, the objective was to make sure that the clock would be in top condition for its big birthday. As soon as 8 a.m. struck, abseilers descended on a dial to commence

FIG 16.1 Rear view of the hour-striking barrel. The bolt heads that secure the cam wheel onto the barrel cap were found to be rubbing on the great wheel. As a temporary measure, the bolt heads were reduced slightly.

FIG **16.2** Escape wheel pinion: note the wear on the trundles.

FIG **16.3** The end of the going barrel. Note the key that is one of three that retain the great wheel on its arbor. During the restoration, it was found that the boss was cracked.

cleaning. Inside the clock room, the team of clockmakers was ready, a synchronous drive unit was engaged and the bolts that connected the clutch unit on the clock released; the dials were now free of the Great Clock and were, for the first time ever, driven electrically.

Since the combined weight of the striking barrel, great wheel, arbor, and winding wheel is around ¾ ton, careful thought had to be paid to how it would be lifted out of the frame. A custom-built scaffold with lifting chain blocks running on a girder was the solution. However, this could not be installed until the clock was stopped and delicate items, such as the escapement, removed.

The dials are normally cleaned every three years by a team of abseilers, complete with good old-fashioned buckets and sponges. Inevitably, repairs would be needed to the odd pane of glass and, where putty had fallen prey to the vicious action of freezing and thawing; this would be replaced with silicone rubber.

Finally, an experienced engineering company was lined up for the mechanical repairs. With a three-ft diameter great wheel and a four-ft long barrel, a seriously large lathe would be needed, along with other heavy machine tools. A wide experience of working on heavy parts was an essential requirement.

DISMANTLING THE GREAT CLOCK

The quarter striking was tied off, the hour-striking hammer line released, and the hour and going trains let to run down until their weights lowered themselves gently onto the pile of protective sandbags at the bottom of the weight shaft. Before the scaffold erection could start, the going train was stripped out, along with the striking train wheels. Striking winding wheels had been taken out a few days before, but the huge fly and the rest of the train were carefully removed. The safety brake also was taken out.

When the two barrels and other parts were removed from the clock, they needed to be lowered to ground level and removed to various

FIG **16.4** Erecting the scaffolding to enable the lifting of heavy parts.

FIG **16.6** The striking barrel being lowered down the stairwell.

workshops. There were several routes to the ground: the weight shaft directly under the clock, the ventilation shaft that runs from the belfry level to the ground, and the stairwell. Of these, the easiest option was the stairwell. A company familiar with lifting and manipulating heavy weights was engaged to install an electric hoist at the top of the stairwell and organize the lowering of all the parts.

Once the scaffolding was assembled, the huge striking barrel was then slowly lifted from its

FIG **16.5** Safely in its cradle, the great wheel and arbor are removed from the hour-striking barrel.

bearings with a couple of chain blocks: it was then eased forward in front of the clock and lowered into a specially designed cradle. Safely at rest, the winding wheel was removed and the great wheel and its arbor slid out. These three items were carefully lowered down the stairwell.

When all the parts were safely removed, the restoration work was divided into three major tasks: the dials with associated motionworks, the escapement, and the barrels with their great wheels.

DIAL MOTIONWORK

Dials were taken out of action one at a time and set to 12 o'clock. The motionworks were serviced, along with the anti-friction rollers that support the hour tubes. There are two sets of these, one pair inside the clock room next to the motionwork and a second set on an outrigger right behind the dial. In the past, some of these rollers had seized up; leading to flats on the surfaces of the rollers, so the opportunity was taken to repair them. A small scaffold tower was

FIG **16.7** Scaffold behind dial to access the hour tube and minute arbor rollers.

used to access the various sets of rollers. One of the shortcomings in the design of the Clock is the support of the minute arbor. Normally on a turret clock, this arbor simply rotates inside a bush in the end of the hour tube. At Westminster, the hour tube has four slots cut out of it and the minute arbor is supported by four anti-friction rollers that are held by two cages bolted to the outside of the hour tube. Unfortunately, the cages are complete rings that can only be removed when the hands have been taken off. Fortunately, minimal work was needed on these bearings.

THE ESCAPEMENT

As a complete opposite to the heavy work needed on the barrels and great wheels, the escapement was much more like conventional benchwork. Whilst the heavy work was going on at the engineering company, the escapement was being rebuilt. A new lantern pinion was made for the

FIG **16.8** Two different times: a rarely seen sight as one of the dials is being serviced.

FIG 16.9 The finished lantern pinion and arbor made for the escape wheel.

FIG 16.10 Edmund Beckett Denison's double three-legged gravity escapement. The assembly was brought to Upton Hall for the turret clock course.

escapement and since its arbor had a whole series of holes, it was decided to make a new one. An examination of the escapement took place at Upton Hall during a turret clock course held in September 2007. The gravity arms seemed to be of recent construction; there were several fixing holes, showing that the locking blocks had been moved or replaced at some time. Indeed the whole escapement did not have the look and feel of the rest of the clock, suggesting that it was a twentieth-century replacement. This fits in with the knowledge that the escapement was replaced by a spare in 1976: the gravity arms appear to have been reconstructed in 1976, while the escape wheel still seemed to be that installed in 1877. New blocks and lifting pins were fitted at the time of the 2007 examination, and flats on the gravity arms and the pendulum rod were all polished out.

CRACKS IN THE GOING BARREL

On the going barrel, there are two end plates, each with bronze bushes: the great wheel and its arbor rotate in the bushes. The great wheel is fixed to its arbor using three slots that correspond with three slots on the arbor. Tapered wedges were hammered into the mating slots to secure the great wheel. Cracks were noticed in the great wheel and a further examination was carried out using dye penetrants and magnetic techniques. The force of the wedges had caused the great wheel boss to crack; these cracks propagated from the sharp roots of the slots. Repairs were made using cold stitches, a technique commonly used in industry to repair cracked castings. Here, a series of holes is drilled using a custom jig, the line of holes being at right angles to the direction of the crack. A dumbbell-shaped lock insert is put into the hole and secured with a resin. The lock is made of a high-strength nickel steel alloy and is much stronger than the parent cast iron. Accurate drawings were made of the going great wheel, so in the unfortunate event of the wheel having to be replaced, the necessary information would be at hand. All the bushes were repaired with new bronzes, but it was decided to make a new arbor, the great

FIG 16.11 Cracks in going great wheel boss.

wheel being secured to the arbor with a single conventional Woodruff key, and the key retained with a collar.

Both of these have bronze bushes, which were found to be worn. Also worn was a thrust washer that had allowed the bolts securing the ratchet wheel to rub on the great wheel. First, the main arbor was skimmed down to remove wear. It was then decided to rebush all the barrel bearings and to reline the split plumber blocks acting as the main bearings. One original bush had rotated over time, so the oil hole in the bush no longer lined up with the oil hole in the great wheel. New bushes were pinned to prevent this happening again. It seems that this was the first time in the clock's history that this bushing work had been needed; indeed it is a great tribute to its maker, E.J. Dent. Clickwork and ratchet teeth were built up with weld and reprofiled where necessary.

THE STRIKING TRAIN

The hour-striking great wheel and barrel assembly has a barrel that is a cast-iron hollow drum. The end plates are of cast iron; one is the winding wheel and the other the ratchet wheel.

WEIGHTS AND WINDING MECHANISM

Dent installed a winding mechanism in 1912. One spur gear on the hour-striking train had become badly worn, so a new one was made.

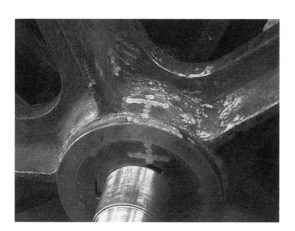

FIG 16.12 Repairs using cold stitches can be clearly seen; these look like a row of filled holes.

FIG 16.13 Bush in end plate of winding wheel: note the oiling holes.

When the weights were on the ground the giant pulleys were all checked and serviced. Lines were inspected and one of the tie-offs was remade. Whilst the work on the clock was going on, other tasks of a non-horological nature that needed doing were completed: the stairwell and handrail was painted once all the parts were back up the tower. All the bulbs illuminating the dials were changed during the maintenance interval.

REINSTALLATION

With the flatbed frame empty, frame members and other parts were cleaned and all the necessary odd jobs completed. As the major parts came back from the engineering works, they were hoisted up the tower and installed; the going train first, then the striking train. First the going train was run without the gravity arms to identify any possible tight spots in the train. The running speed was regulated by the fly on the reconstructed escape wheel. After several fault-free runs, the gravity arms were installed and the clock set going. After a short series of adjustments, the clock was brought to time and remained under test for a week before the official restart.

Once the strike barrel had been returned, it was installed and tested. The train and winding wheels were reinstalled, along with all the release and locking levers. The scaffolding was then struck and the striking fly and winding wheels reinstated. After a period of testing, the Great Clock was deemed fit to resume normal duty.

On the evening of Saturday, 29 September, the bolts on the main hand-setting clutch were tightened and the synchronous motor disengaged. The Great Clock was again running as its maker and designer had intended. Over the next day, the timekeeping performance was spot on.

Monday, 1 October was the official restarting of the clock: Just before midday, the tied-up quarter and hour trains were released, and shortly afterwards the Great Clock sounded the Westminster quarters and the twelve hours, after a silence of seven weeks. The project was completed; the rails and barrier were replaced and all loose ends tidied. A few weeks later, the official tower guides were back in business, taking their tours up the tower every morning.

FIG 16.14 The new spur gear on the motor-powered winding mechanism.

THE DRIVE UNIT

A drive until was installed to drive the dials during the period of the restoration. The requirement was that the unit had to drive the four dials, and it had to be easily engaged and disengaged. An inspection of the bevel gear cluster above the clock revealed that space was rather limited but fortunately the bevels were secured to their arbors by keys providing something secure with which to make a connection.

The Cumbria Clock Company, which specializes in turret clockwork, was commissioned to

the opposite arbor. Since there was a nest of bevels in between them, the sprockets would rotate in opposite directions so a counter-rotating drive was needed.

A solid aluminium base and case was built for the drive unit: the case held the various arbors, which run in commercially available flanged sealed ball races. A heavy-duty synchronous motor provided a 1 r.p.m. output; this was connected to a hardened steel worm giving a 60:1 reduction in conjunction with a bronze worm wheel. Chain sprockets were mounted on the appropriate output arbors. To engage and disengage the drive, the worm and motor were mounted on a swinging frame, the position of which could be selected to drive or not to drive.

FIG 16.15 The great clock's leading off on the girders above the clock. The A-shaped frame is not part of the bevel gears; it is a lifting winch known as a double-barrelled crab. The drive to the dials is through the diagonal rod from the clock to the left-hand bevel cluster. The drive then goes to the other side of the crab, where it is split by more bevels into three drives for the other dials.

FIG 16.16 Under test, and lifting two weights totalling 2 cwt.

make the drive unit, which was designed by Keith Scobie Youngs, the managing director. The final design comprised two chain sprockets that are split, enabling them to be clamped with bolts onto the bevel gear arbors without having to dismantle anything. These were computer-designed and laser-cut for accuracy. A slot in the sprocket arbor engaged with the projecting head of the securing key on the bevel gear, providing a positive non-slip drive. Two sprockets were decided upon; one on one arbor, the other on

FIG 16.17 The synchronous drive's traditional setting dial.

A pre-settable slipping clutch was incorporated to prevent damage in the unlikely event of something jamming. A nice finishing touch reflecting the company's passion for traditional clock-making was a hand-engraved silvered setting dial.

Under test, the drive unit was securely clamped and turret clock weights added to both arbors to simulate the worst-case scenario. The drive unit purred away and easily coped with the full-load test situation.

CHAPTER SEVENTEEN
THE GREAT CLOCK—A COMPLETE DESCRIPTION

THE FRAME AND CONSTRUCTION

The Great Clock sits on two piers, which are the north and south walls of the weight shaft. The clock frame comprises two main members: the back and the front. Three crosspieces hold the main frames together, one at each end, and the third in the middle. All frame members are made of cast iron and painted black. The whole clock is constructed on the flatbed principle, so that any wheel may be removed without disturbing another or having to dismantle other parts.

With one exception, all the wheels are made of cast iron, the teeth being part of the casting and not cut afterwards. Denison claimed that this was the best method for cheapness and strength. He also advocated that all non-working surfaces should be painted. The tradition in the clockmaking trade was often to highly polish parts, even those that did not need it like the side faces of wheels. Denison dismissed this practice with all his venom as '*Working for fools*'.

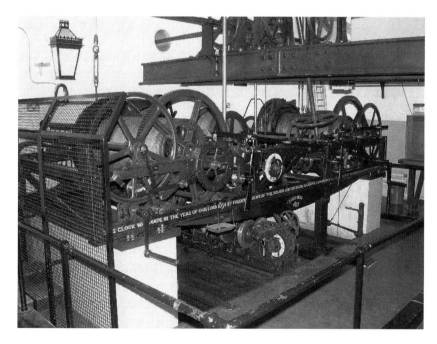

FIG 17.1 The Great Clock.

227

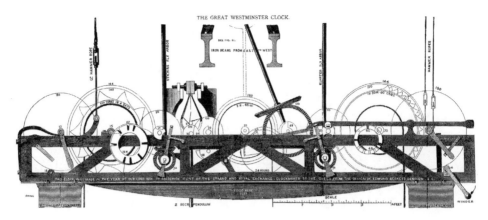

FIG 17.2 The Great Clock as depicted from 1860 onwards in *A Rudimentary Treatise on Clocks, Watches, and Bells.*

THE GOING TRAIN

Central to the going train is the great wheel with its barrel; when fully wound this will drive the clock for ten days. The barrel is a wrought-iron tube, to which is fitted a winding wheel. The great wheel drives four separate motions; the going train to the escapement, the 24-hour wheel and two one-hour arbors. One of these lets off the quarters; the other lets off the hour and drives the dials through the bevel gears.

Precision clocks need a maintaining power mechanism to keep the clock running whilst it is being wound. The Great Clock has an unusual and effective device that has only been used at Westminster. To wind the going train, the winding pinion is first slid out into engagement; next a wind indicator is pulled out. This is simply a long rod with a projection. When the going train line traverses the barrel and reaches the fully wound point, the rod is released, signifying that the clock is fully wound. The engaged winding pinion is on a swinging arm; this arm has a click that engages with flat ratchet teeth set on the rear face of

FIG 17.3 A general view of the going train, showing the barrel and the great and winding wheels. The hour arbor meshing with the great wheel drives the hands; the internal setting dial is on the end.

the great wheel. As the pinion is wound, the winding wheel turns; the reaction is via the click on the flat teeth. In this manner, the maintaining power is the same as the normal applied force. As the going train is wound, the great wheel continues to turn and the swinging arm with the pinion is carried round with the turning wheel. A stop arm on the pinion eventually contacts a stop piece and the person winding has to stop and disengage the winding pinion, so

FIG **17.4** The right-hand hour arbor that releases the quarter-striking train.

FIG **17.6** The setting dial on the rear of the hour arbor that drives the hands. On the large pinion is a nut, which, when loosened, releases the drive to the dials.

FIG **17.5** The right-hand hour arbor that releases the hour-striking train. The four-lobed cam was once part of the quarter-striking barring off mechanism.

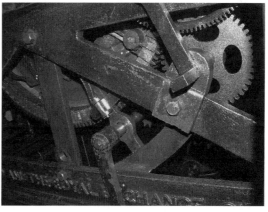

FIG **17.7** Maintaining power. The swinging arm hands at about 5 o'clock. The click engages with the ratchet teeth on the face of the great wheel. The pinion on the right is a remnant of the unused electrical winding unit. In front is the stop for the arm on the winding arbor.

that the swinging arm returns to its start position. Then the winding can begin again.

The going train comprises the second wheel, which turns in 15 minutes. On this wheel is a cam, which is used as part of the precision let-off for the hour striking. Two seconds hands are fitted onto the third wheel; one can be viewed from the front of the clock, the other from the rear. Since this wheel turns once in two minutes, the indicator dials are graduated from 0 to 60 twice. The escape wheel is of the double three-legged type and the fly has three blades. Most turret clock flies have only two blades, which are easier to make and to balance.

FIG **17.8** The dual seconds dial, as viewed from the rear, with the 15-minute cam to the left.

FIG **17.10** The drive to the 24-hour wheel is through a pinion of report on the barrel arbor. The notches in the 24-hour disc can just be seen in the bottom right-hand side.

FIG **17.9** The gravity escapement. The second wheel and third wheel can be clearly seen, as well as the maintaining power ratchet.

FIG **17.11** The 24-hour dial. Again the notches in the 24-hour disc can be seen. The lever running from bottom left to top right is the release lever for the quarter striking.

The 24-hour wheel once closed contacts that were part of the system where the Great Clock signalled its time to Greenwich Observatory. It may have been intended to be used as controller for turning on and off the gas lighting, though this was never used.

Going Train Details

Wheel	Teeth	Pinion	Revolutions per hour
Pendulum			1,800 beats per hour
Escape wheel	6	9	300
Third wheel (carries a seconds hand that traverses a dial of 2 minutes)	90	16	30
Second wheel (carries a 15 minute precision let-off cam and a minute hand that traverses a setting dial)	120	12	4
Left-hand hour wheel (setting dial and quarter let-off)	48 (bevel gear drive to dials: 120)		1
Right-hand hour wheel (drive to hands and hour let-off)	48		1
Clicks	3		
Ratchet wheel	36		
Great wheel	180	20	4/15 = 0.2666 (once every 3 hours 45 minutes)
24-hour wheel	128		1/24 (1 turn in 24 hours)
Barrel	No grooves but it takes about 60 turns		
Barrel-winding wheel			
Winding pinion			
Driving weight	5 cwt		
Fall	200 feet		

THE PENDULUM

The pendulum is of the compensation type, invented by Dent and first used around 1849. Concentric tubes of iron and zinc compensate for temperature changes. The bob is cast iron and weighs about 450 lb. Dismantling and photographing the pendulum is not an option, so the following photos are of a similar pendulum and are intended to show the principle.

Pendulums are normally regulated by raising or lowering the pendulum bob using a nut at the bottom. As it is very inconvenient to raise or lower such a massive bob, regulation is achieved by adding or tasking away small weights. Near the top of the pendulum is the collar forming the upper end of a steel tube, part of the compensation system. An old penny added to the collar would cause the clock to gain 2/5 of a second a day. To get the clock to time in two second increments is easy; either the escapement is held up for one tick or it is allowed to advance by an extra tick. To adjust the time by one second, two shifter weights of around 5 lb were supplied. When one

FIG **17.12** A compensation pendulum with the bob removed. The crosspiece near the top is used for regulating weights.

FIG **17.13** The dismantled pendulum rod. From top to bottom: steel rod, steel tube, zinc tube.

FIG **17.14** The regulation weights. The large U-shaped weight underneath is the one used for making large corrections quickly. Its counterpart, which was added to cause a gain, is now lost.

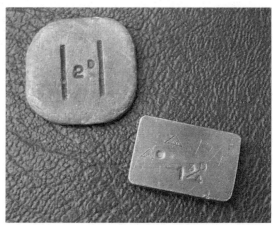

FIG **17.15** Small regulation weights. One carries Dent's trademark.

was added to the regulation collar, the clock would gain one second in 10 minutes. If the shifter that was permanently on the collar was removed, the clock would lose a second in 10 minutes.

A massive cast-iron bracket is let into the air shaft wall. This supports the pendulum. Only the top of the pendulum is visible in the clock room; the main part and the bob swing in a specially constructed iron box below the clock. At the bottom of the pendulum a beat plate is fixed to the wall to enable the pendulum arc to be measured, in degrees. Underneath the bob are two large nuts that allow the bob to be lowered or raised for large-scale adjustment. Stored in the pendulum pit are two spanners specially made for adjusting these nuts.

THE HOUR-STRIKING TRAIN

The hour-striking great wheel carries ten cams, which operate the bell hammer: a weight of

FIG **17.16** The hour-striking train.

FIG **17.17** A rear view, showing the steel-faced hammer-lifting cams.

FIG **17.18** The locking arm is fixed on the end of the arbor that drives the fly, it rotates anticlockwise. It is shown here in its normally locked position. Note the second stud just below the top end of the arm. This touches the L-shaped lever on the right-hand side and prevents the train from running backwards when it is being wound. The hook end of the lever can be used to stop the striking for maintenance purposes.

FIG **17.19** The horizontal locking lever has been raised by the hour cam and released the locking arm to its second locked position, the warning position.

FIG **17.20** After the hour cam has dropped, the locking lever is now in its final warning position, waiting to be released by the 15-minute cam.

around 1 ton provides the motive power. Speed regulation is by the vertical fly, which is driven through bevel gears, and a conventional count wheel controls the number of blows sounded. It is unusual, if not unique, that there is not a wheel in the striking train that turns once for each stroke sounded. There is instead a wheel that performs two-thirds of a turn and, as a consequence, the locking lever lifting cam has two notches 120° apart.

Since the specification for the clock demanded that the first blow of the Great Bell to be accurate to a second of time, there are some major differences between the Great Clock and a conventional turret clock. The first difference is the let-off; in a normal turret clock there is a cam on the hour arbor that releases the striking, but the accuracy is probably only to within 15 seconds, owing to backlash in gearing. A two-stage release is used in the Great Clock. A cam on an hour arbor raises the locking lever and a few minutes before the hour this releases the

FIG **17.21** Caught in the action of striking the locking arm is rotating. Note the L-shaped lever on the right of the picture. This acts as a click to stop the train turning backwards when it is being wound and the hook enables the striking to be disabled for maintenance purposes.

FIG **17.22** The 15-minute cam shown just after it has dropped the final locking lever.

FIG **17.23** 'On the lift' and well clear of the bell, the hammer is poised to make its first blow. Note the two round rubber pads that keep the hammer head just clear of the bell once it has struck.

locking arm, which was held on a steel block. The arm turns a small amount and locks on a second block, known as the warning. In a conventional clock, the striking starts when the locking lever falls off the cam, but in the Great Clock the locking arm advances to a third stop that is held by the 15-minute cam, which is on the second wheel of the going train. The cam is always advancing, there is no backlash from the pressure of driving the escapement, and the edge of the cam is faced with a hard steel strip. This means that the drop-off can be given to the second. The cam raises the final locking lever during the last 15 minutes before the hour and drops it two seconds before the hour. This final action releases the train and the bell is struck.

When a turret clock strikes normally, the train runs, lifts the bell hammer, and lets it drop, a process that takes perhaps six seconds and might vary as the train speeds up. In the Great Clock, the bell hammer is *left on the lift*, i.e., it is fully raised, and when the hammer lifting cam turns a fraction more, the hammer is released and sounds the hour. By operating the hammer in this manner, the variables are virtually removed and the bell is sounded precisely on the hour.

Hour-Striking Train Details

Wheel	Teeth	Pinion	Revolutions per stroke
Fly	72 (bevel gear)		4
Locking lever arbor	72 (bevel gear)	15	4
Second wheel	90	21	⅔
Count wheel	117		1/78
Pinion of report	15		
Great wheel	140		1/10
Cams	10		
Barrel	64 grooves		62.5 turns in four days
Clicks	8		
Ratchet wheel	44		
Barrel-winding wheel	144		
Reduction winding wheel	150	12	
Winding pinion		14	
Driving weight	1 ton		
Fall	200 ft		

FIG **17.24** The quarter-striking train.

FIG **17.25** The quarter barrel, showing the lifting cams.

THE QUARTER-STRIKING TRAIN

In construction and operation, the quarter-striking train is similar to the hour-striking train. On the great wheel are five circles of hammer lifting cams. There are five sets of cams, since the fourth quarter bell is required to strike twice in succession and it is easier to provide two hammers rather than make one hammer work quickly. Each quarter chime is a phrase of four notes and there are a total of ten phrases in an hour. The great wheel turns once in 1½ hours so there are 15 phrases in all, giving a total of 60 cams.

A count wheel controls the striking. This has three sets of notches, so it turns once in three hours.

Quarter-Striking Train Details

LEADING-OFF WORK AND MOTIONWORK

The hands on the dials are connected to the clock by a rod that turns once an hour, this rod comes off the right-hand hour arbor and the motion is transmitted through bevel gears. From the clock, the rod passes up at an angle to the girders that run east–west above the clock; here the rod connects with another set of bevel gears. The motion is sent to the east dial and a rod heads west, where it drives another set of bevel gears. From this set the drive goes to the north, south, and west dials.

Behind each dial is a set of motionworks, this has a 12:1 reduction gearing that drives the hour hand to produces one turn in 12 hours. The minute hand is mounted on a long arbor, and the hour hand is fixed on a tube concentric

Wheel	Teeth	Pinion	Revolutions per four-note phrase
Fly	72 (bevel gear)		3
Locking lever arbor	72 (bevel gear)	15	3w
Second wheel	90	20	
Countwheel	60		1/3 per hour
Pinion of report	30		
Great wheel	150		2/3 turn per hour: one turn in 1½ hours
Cams	60		
Barrel	64 grooves		64 turns in four days
Barrel-winding wheel	144		
Reduction-winding wheel	150	12	
Winding pinion		14	
Driving weight	1 ton		
Fall	200 ft		

FIG **17.26** Bevel gears above the clock route the motion to the four dials.

FIG **17.28** The cast-iron outrigger that supports the weight of the hands.

FIG **17.27** One of the sets of motionworks.

FIG **17.29** Roller bearings; the large set takes the whole weight of the hands; the small set supports the minute hand arbor.

with the minute hand arbor. Antifriction rollers support the weight of the hands. The hands must be carefully counterbalanced so that all the clock has to do is to turn them. Both hands have small counterbalance weights in the tails of the hands but internal weights are provided to complete the job. For the minute hand, these weights are supported on a Y-shaped arm; using two weights allows for an accurate balance to be achieved—this cannot always be done using a single weight. A single weight counterbalances the hour hand but this is difficult to see from the floor of the clock room.

Behind the dials, the weight of the hands is supported by an outrigger projecting from the tower wall. At the end of this is a set of supporting rollers. A cage of four rollers is built into the end of the hour tube, and provides the outer bearing for the minute arbor.

TIMING THE GREAT CLOCK AND CONNECTION WITH GREENWICH

After Airy saw Shepherd's electrical clock at the Great Exhibition of 1851, he saw just how useful this could be. He soon installed two Shepherd master clocks with sympathetic or slave dials distributed around the Observatory. One master clock told sidereal (star) time; the other mean time. A large dial was put up outside the Observatory gates for the use of the public and, over a period of time, telegraph lines were run to the Post Office, which distributed time signals. Other lines sent a time signal to operate the time balls at Deal and Start Point near Plymouth.

Every year, the Astronomer Royal produced a report on the Royal Greenwich Observatory's activities. *The Report of the Astronomer Royal to the Board of Visitors* fell into a set format that

FIG **17.30** The master clock by Shepherd that Airy had installed at Greenwich after 1851.

summarized the buildings, equipment, library, observations, publications, staff, and time service. The Board of Visitors went to the Observatory in June each year, and the report covered the period up to May of that year. In the 1863 report, Airy wrote,

The Clock of Westminster has been brought into connexion and the attendant receiving a signal from us every hour and the clock reporting its state to us twice a day. As far as I have observed the rate of this clock may be considered to be much less than a second per week.

A direct-line telegraph went from Greenwich Observatory by overhead lines alongside the South Eastern railway line to London Bridge station. From there it went across London Bridge to the Electric Telegraph Company's head office at Founders Court, in the City. A line ran from the city along to Charing Cross via Fleet Street and the Strand. From Charing Cross the line ran to the Houses of Parliament via Trafalgar Square, Whitehall, and, finally, Parliament Street. Dent's premises were in The Strand, where they had an instrument to receive the time signal and, presumably, check the clock's performance.

In the following years, a fuller report of the Great Clock's performance was given. Generally, the report took the format of a percentage, giving the number of days when the clock was accurate to within one second, between one and two seconds, between two and three seconds and between three and four seconds. In the period from 1860 to 1890, the telegraph was still an emerging technology, experiencing reliability problems due to the continuity of the line, its insulation, and the condition of batteries. Sometimes the number of days when no signal was received from Westminster was reported: 1890 was particularly bad, with 44 days of no signal, while 1876 was a particularly good year, when the clock was within one second for

273 days. However, the following year, 1877, Airy stated that, '*The Westminster Clock from some neglect of the chronometer maker has not fully maintained its character it has sometimes been more than 3 seconds in error.*' By 1878, he was able to say,

The Westminster Clock was cleaned by E. Dent & Co. last autumn, the clock being out of use from 21 August to 11 October. Since the cleaning the clock has gone with a remarkably steady rate though sometimes affected by high wind. During the period to which the report refers the West-minster Clock has been 3 seconds in error on 7 days, 2 seconds on 21 days and for the rest the error was under 1 second.

In 1886, Christie, Airy's successor, reported that, '*The contact apparatus was out of order since no signal had been received in the last 12 months. A new contact apparatus is now being fitted.*'

Greenwich reported the Great Clock's rate for over 80 years but the annual report for 1941 recorded that, '*Since September 1940 it has not been possible to maintain a check on the Westminster Clock.*' Enemy action in World War II destroyed the telegraph link.

Since the manner in which the figures are presented is not consistent, an analysis is best divided into pre-1910 and post-1910. Before 1910, the clock was on average within 2 seconds of the correct time for 84% of the days measured. After 1910, the time was correct to within a second for 81% of the days when measurements were taken. However the figures are viewed, this represents an astonishingly good performance for a mechanical clock.

The Great Clock automatically reported its time to Greenwich Observatory twice a day. Undoubtedly, one of the times was 1 p.m., when time balls were dropped. A drum chronograph was used to record the signals from the West-minster Clock, along with signals received from various time balls and other clocks. The chrono-graph comprised a paper chart wound round a drum that was driven by clockwork at a constant speed. Pens drew lines on the paper chart; the pens were moved by a telegraphed current when an event took place. Timing marks from the observatory standard clock were also recorded. Examination of the chart meant that an event could be measured to within 0.1 of a second. The second time that the Westminster clock sent its signal to Greenwich was probably in the evening, as the astronomers were setting up for their night's work.

Photographs of around 1907 and Dent's promotional drawing of the Great Clock show a contact and two wires situated just inside the clock frame below the 24-hour dial. This must have been a switch that was closed for the two times in the day when the signal was to be sent to Greenwich Observatory. Two arms fitted to the 24-hour wheel at appropriate places would then close the switch. There are notches on a disc on the 24-hour arbor and these may have held moveable arms that could be set to the desired hour for signalling to Greenwich. To signal the exact instant of Beg Ben striking, a switch would have had to have been fitted adja-cent to the hammer-lifting arm; this would be closed at the instant the lifting arm fell. There is no trace of the switch or of where it might have been fitted. Since the telegraph line was a single wire, some means would have been needed to switch the line between a galvanometer that received the time signal from Greenwich and the switch on the clock that sent its hour signal back to Greenwich. On the hour arbor of the going train there is an arm that has an agate jewelled slip in the end and there are some unused holes on the frame adjacent to the arm. Certainly, this was once an arrangement used for operating switch contacts; probably this set changed the line over from receive to transmit just before the

FIG 17.31 An assistant inspecting a chronograph drum at Greenwich. The chronograph is the instrument to his left in the case.

hour and returned the line to receive just after the hour. In this proposed mode of operation, the hourly time signal from Greenwich would have had to be sent at some time other than on the hour. In the correspondence leading up to the ordering of the Great Clock, Airy mentioned a time signal on the half hour. The clock's time would be checked by reading the seconds dial and comparing it with the needle on a galvanometer or telegraph instrument. The galvanometer must have been next to the setting dial, where it would be easily read. No photos have yet been found that show the galvanometer.

Since the clock was wound three times a week, there were ample opportunities to adjust the clock if it was wrong. On this basis, one might question that the observations done at Greenwich were only ever over a 2 day period, after which the clock might have been corrected. Restoration of the telegraph link between Greenwich and Westminster after the enemy action in 1940 was a low priority. It seems that after the war it was agreed that the Great Clock had proved itself and no longer needed to be monitored by the Royal Observatory.

Timing the Great Clock the Modern Way

In the period from 18 March 1977 to 2 June 1978, a period of 350 days, Michael Maltin logged the

Great Clock's performance. The check was done by comparing the first stroke of Big Ben received on the radio against the MSF time signal from Rugby. Both signals were captured on the display of a storage oscilloscope and the time difference between them measured from the screen. The clock was within ±1 second of the time signal for 88% of the period and within ±1.5 second for 94%. There was a period of 15 consecutive days when the clock was within ±0.1 second.

Maltin logged his observations in graph form and observed how the Great Clock's time was coaxed to be correct for Remembrance Sunday, 13 November. The clock was 1.5 seconds fast on the previous Monday but by Sunday it was 0.3 of a second slow.

Less successful management was observed on New Year's Eve, to welcome in 1978. Leap seconds are sometimes added to civil time and this is normally done at the end of the year, making the year a second longer. If Big Ben was not put a second slow, then the New Year chimes would sound one second early, since the year had been extended by a second. Maltin's recording showed that in the run-up to 31 December the clock was persuaded into being 1.5 seconds fast; no doubt, with the intention that when adding a leap second, the clock would be only half a second out. However, a simple error had been made, and the clock should have been set to be 1.5 seconds slow, not fast. The net result was that the New Year of 1978, as heralded by the Great Clock, was 2.5 seconds early! Within six days, the clock was back to a near zero error.

Checking the Great Clock in the Twenty-First Century

When the clock is wound, a stopwatch is used to measure the difference between the first stroke of the hour and the telephone speaking clock. There is a phone line in the clock room and the time is checked shortly before the clock strikes. A record is kept of the Great Clock's performance and adjustments are made when necessary.

A sophisticated commercial clock timer has been connected to the clock and this revealed an astonishingly level rate. However, a large transient deviation of several seconds was once observed that lasted about an hour. It is believed that this coincided with high winds. An estimate given by an engineering lecturer indicated that a tall building like the Clock Tower might sway one or two inches under the effect of wind, with a period of five to ten seconds. Such a movement would certainly affect the going of the pendulum.

WINDING THE CLOCK

Once the clock had been installed, winding it soon caused controversy. It took two men three days a week full time to wind the clock, probably about 48 man-hours. Compare this with a large church clock of the time, which would take, at the most, an hour to wind, and the huge disparity is obvious.

There were proposals to introduce an automatic form of winding; one idea was to harness the tide in the Thames, while another was to use spring-loaded platforms in Westminster Bridge that would be actuated by people walking over them. In June 1859, Jabez James, the engineer who was responsible for raising the bells, applied for a patent that used a special version of a steam engine to raise clock weights. From the date of the application, it seems that James saw the problem with winding the Great Clock and opportunistically filed to patent his solution.

WINDING UP THE CLOCK

"BIG BEN" AND THE CLOCK TOWER, WESTMINSTER PALACE

FIG **17.32** Men stripped to the waist demonstrate the tedious nature of winding the Great Clock, from *The Graphic* of October 1887.

His idea used a steam piston, which, as it reciprocated, used a system of pawls to pull a chain in one direction only. The invention removed the need for a crank. James went on to propose that his invention could be used as the driving weight of the clock, the engine climbing up the chain that was connected to the clock and the steam being provided through flexible hoses. The inventor was over 100 years ahead of his time. It was not until the 1970s that Thwaites and Reed produced an automatic winder for turret clocks, dubbed a *monkey-up-a-rope*. Here an electric motor with integral gearbox climbed up a chain connected to a sprocket on the clock, the weight of the motor-gearbox providing the motive power.

The most sensible idea for winding came, of course, from Edmund Beckett Denison, who proposed a hydraulic ram under each weight. The weight line would make several turns round the barrel and have a smaller weight on the end. At an appropriate time the ram would rise and lift the main weight, the barrel would 'wind up' a few turns, and then the ram would lower to its rest position, leaving the weight wound up sufficient for the next hour's work.

Rudolph de Cordova wrote an article for an American magazine, in which he accurately describes the action of winding the Great Clock. Although this appeared in 1901, the description must have been as true as when the clock first started.

To wind the clock two men are engaged three afternoons every week. The going part of the machinery is wound in 20 minutes, but it takes five hours to wind the striking. The winding of the striking is by the turning the handles of two winches connected with the barrels on which is wound the wire rope which carries the weight which drives the machinery, 125 turns on the handle being necessary to get one turn of the barrel. And as there are 60 turns of rope on each barrel each man has to make considerably more than 7,000 turns of the handle to wind his part of the machinery.

It is no easy work winding Big Ben, and the men have to strip naked to the waist for the purpose, while they have to rest at frequent intervals. As they turn the handles there is an incessant clanking of the machinery which sounds for all the world like as if one were in a forge, the noise being made by the clocks dropping onto the ratchets, the notches of which are a good inch or more deep.

As each quarter approaches the winders are compelled to leave off so that the machinery may be quiet for the clock to strike. The men need not however, watch the time for this, for, before it strikes, the clock itself always gives a special warning, one for the quarters and one for the hours.

To wind the striking trains, the winding gear has first to be engaged; this is done by turning an eccentric bearing that brings the pinion on the winding wheel reduction gear into engagement with the winding wheel on the barrel. The double-stage reduction means that just over 128 turns on the winding handle are needed to produce just one turn on the barrel, raising the weight of approximately 1 ton by around three feet. To wind the striking trains completely takes 8,100 turns on each winding handle.

A report in *The Engineer* of 1856 stated that springs were used to lessen the noise made by the clicks as they fell. Obviously, such a contrivance was removed at some time in the past.

Winding Barring

It is not allowable to wind the striking trains when the clock is in the process of striking, since the action of winding removes power from the trains and thus the clock would not strike. It would be easy to overlook having a break just before a quarter struck, so a barring mechanism was installed to prevent winding at the critical time. The 1860 drawing of the Great Clock in *A Rudimentary Treatise on Clocks, Watches, and Bells* shows a simple lever on the quarter train that is raised by a four-lobed cam on the right-hand hour arbor. As this is lifted, it finally engages with an arm on the winding-handle pinion and thus prevents further winding; a spring is

FIG **17.33** The stopping arm with its spring-loaded tail; a relic of the first winding-barring mechanism.

FIG **17.34** The winding-barring mechanism on the hour-striking train.

included in the arm to cushion the shock of stopping. The operator then has to throw the winding gear out of action by moving the eccentric bearing. Once the clock has finished striking the quarter, the winding can again resume. This mechanism was replaced by a more sophisticated device when the clock was installed in the Clock Tower. The sprung-loaded stopping arm on the winding-handle pinion can still be seen.

The first barring mechanism was replaced by a more complicated successor. Probably the slow lifting of the arm that disabled the winding meant that the stopping was not a precise operation. In the second version, a barring lever is first raised and held up by a catch. A lifting lever is raised by the quarter cam and this eventually released the barring lever that falls and prevents further winding, by engaging with an arm on the winding handle. Denison wrote that this mechanism was not really needed on the hour-striking train since the sounding of the last quarter was sufficient warning.

Today, the barring mechanism on the quarter-striking train is missing. Photos of the clock appeared in the May and July edition of the *Horological Journal* for 1959 showing the quarter barring mechanism still in place; it is most likely that the mechanism was removed when the quarter-striking train was rebuilt in 1976. On the hour-striking train, most of the barring mechanism remains but the locking arm on the winding handle has gone. A lifting lever is now also missing. Older photos show the lever that runs inside the frame pivoted on an arbor just behind the 24-hour wheel. The right-hand end extends to the right-hand hour arbor and engages with a cam that is still present. It is not known when the hour barring mechanism was disabled.

The Electric Winding Unit

An electric winding unit was designed, manufactured, and installed by Dent in 1912. The unit is situated on the floor under the going train and comprises a large 5-horsepower electric motor that drives a horizontal shaft. Three friction clutch units on this shaft drive roller chains when engaged. One drive is for winding the hour-striking train, one for the quarter-striking train, and one the going train. Beneath the hour-striking train, the roller chain runs to a two-stage set of reduction spur gears, and then through a dog clutch to a pinion that engages

FIG **17.35** The electric winder with the motor at the rear and friction clutch units. At the front is the controller unit.

with the winding wheel. In normal operation, the winding wheel drive pinion is disengaged, so the clock does not drive the winding gear when it is striking the hour. When the striking train needs to be wound, the friction and dog clutches are engaged, connecting the drive through to the winding wheel. The same system operates for the quarter-striking train.

Since the full winding of the clock takes around 40 minutes some system was necessary to prevent the winding taking place when the quarters and hours were striking. This was achieved by having a cam wheel on the hour arbor; this was made in two halves so that it could be clamped onto the arbor without having to dismantle part of the clock. The cam wheel turns once an hour and the cams lift a link lever that connects with the mechanical controller mechanism mounted below on the front of the electric winding unit. There are three cams at each 15-minute space. About 1½ minutes before striking the quarter commences, the first cam lifts the connecting rod to the controller; this causes disengagement of the friction clutch and

FIG **17.37** Under the quarter-striking train, the chain drive from the friction clutch runs to reduction gears. The dog clutch can just be seen in the top left-hand corner.

FIG **17.38** The striking train overwind switch in the set position.

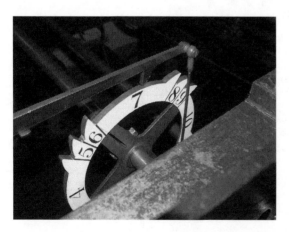

FIG **17.36** Cams fitted on the arbor signal the electric winding to turn off, to allow the quarters and hour to strike.

removes the power to the winding. The winding wheel then runs backwards a little. This happens slowly because of the friction in the spur gear cluster and has the effect of lowering the clicks gently onto the ratchet. The second cam on the trio then lifts the connecting rod to the controller and this causes the dog clutch to be disengaged, allowing the winding wheel to run freely as the clock strikes. The clock then strikes and, shortly after it has finished, the third cam lifts the connecting rod; this time the friction and dog clutches are re-engaged and winding commences. On the hour, a fourth cam has the effect of re-engaging the hour striking. When the cam wheel operates the link, the spring-loaded controller wheel is advanced by an escapement mechanism. Cams on the controller wheel operate the various friction and dog clutches.

To prevent overwinding, a safety cut-out mechanism is employed that disengages the friction clutch. As the wire line is wound along the barrel, it traverses the length of the barrel. A hinged flap has been positioned on the clock pier, so that if the wire traverses too far, the flap is pushed to one side; this then releases a weighted arm that cuts out the friction clutch. Overwind switches are fitted to both the quarter- and hour-striking trains and these are set before winding starts.

The person winding the clock has first to unhook the chain that secures the connecting rod and allow it to fall into one of the numbered slots in the cam wheel. The next step is to set the controller wheel on the electric winder unit to the same number—this ensures that the proper sequence is observed. The operator has to wind up the controller spring by disengaging a wheel, giving it a turn, and re-engaging it. The overwind catches are then released and finally the power to the motor is turned on and winding begins.

When winding has been completed, the connecting rod is hooked up out of action and the motor turned off.

It seems that the winding mechanism for the going train was unsuccessful since its components have been removed and are now stored in the tunnel to the west dial. A photo of the Great Clock in Alfred Gillgrass's book shows the motor winder. The chain drive to the going train can be seen, so winding the going train was still done electrically in 1946 when the book was written. Today, the going train is wound by hand and this only takes about 20 minutes. There were a few incidents in the 1930s when the motor winding mechanism failed and stopped the clock; the worst case was when it caused a tooth to be broken off the striking train. Today, the winding is stopped manually before the quarters and the hour strikes to ensure that there is no chance of such a problem happening again.

Vaudrey Mercer, in his book on E.J. Dent, reproduces Dent's instructions for using the electric winding. These are somewhat brief and were probably intended as a reminder to the clock winder, who was familiar with the winding operation.

- *Switch on current at bottom of tower.*
- *Switch off dial lights from room below clock room.*
- *Wait until lever would drop into one of the long spaces, then unhook, free pinion on automatic and set to same number as on snail, and also set spring by pulling wheel round towards you.*
- *Pull main switch and then motor switch gradually, pull up the quarter & hour weights by the chains.*
- *To start the going winding, put winding pinion into gear, also maintaining pinion, start by pulling lever forward to take up drive which will be on No. 10.*

After winding:

- *Switch off main switch only and wait till motor stops, hook up lever by chain and leave automatic set to Dent, put winding pinion out of gear, also maintaining pinion.*
- *Switch off light from room below.*

(*signed*) E.B. (*Ernest Buckney*)

ALTERING THE CLOCK FOR SUMMER AND WINTER TIME

Clocks are officially put forward or back at the ends of March and October; this happens at 2 a.m., to cause minimal disruption to the public. For the Great Clock, this is normally done in conjunction with scheduled maintenance and takes place from around midnight to 4 a.m. In preparation for the work, the dial lights are first turned off.

To put the clock forward an hour, the quarter chimes are disabled just after the third quarter has struck. This is achieved by using the anti-reversing click, which doubles up as a striking arrestor. A hook on the end of the click engages with the locking piece, preventing the train from running when it should have been released by the locking lever. Next, the great bell has to be silenced by pulling off the hour hammer. A long turnbuckle is attached to the hammer lifting lever and, as it is tightened, the lifting lever is pulled clear of the operating cams. The last stage is to allow the going train to run free, which is done by tying up the two gravity arms. The clock runs but at an acceptable speed controlled by the fly on the escape wheel.

When the hour-striking train is released, there is a danger of overspeeding and tripping out the safety brake because there is no load in lifting Big Ben's hammer. To prevent problems, the fly arbor is restrained with a gloved hand and allowed to run gently at an acceptable speed.

When the clock has been advanced by an hour (this takes about ten minutes), the going train is stopped and the gravity arms re-engaged and the clock brought to the exact time by letting the escapement run through or holding it up through a few seconds.

To recommence operation, the quarter-striking locking piece is disengaged from the arrestor hook just before the fourth quarter is to sound and the turnbuckle is then released to enable hour striking to take place. The clock is then ready to strike.

To put the clock backward an hour, the clock is simply stopped for one hour just before the hour is due to strike. Stopping the clock is very simple and merely involves tying up the escape wheel to stop it turning. An hour after the clock has been stopped, the tied-up escape wheel is released and the clock carries on ticking. The pendulum is not stopped during this hour, indeed its arc would have only dropped a little and it still is able to operate the escapement.

Once the clock has been put forward or back and the maintenance completed, the dial lights are turned back on and the clock is again proclaiming the time to the public.

DOUBLE-BARRELLED CRAB

Mounted above the clock on the beams that carry the leading-off work is a double-barrelled crab. This is a lifting winch that was used to haul all the clock parts up the weight shaft into the clock room. Edmund Denison showed a small version of this to the Institution of Civil Engineers in February 1859. In the report of the meeting, it was implied that Denison presented the crab as a new invention. It was, however, said that the machinery maker, Richard Roberts of Manchester, had indeed invented this device some 30 years earlier. In his characteristic style, Denison then replied to say that it was surprising that such a useful device was unknown to the builders and engineers to whom it had been shown.

The double-barrelled crab is made up of two grooved barrels with rope passing from one to the other. Both barrels have winding wheels that are driven by a hand-cranked pinion. Friction keeps the rope tight on the barrels and it has the advantage that there is no build-up of layers of rope on the barrels or need to move the rope back once it has traversed the barrel, as is done with a conventional lifting crab.

MEASUREMENTS OF THE GREAT CLOCK

In 1976, on a day when the clock was stopped and undergoing non-destructive tests, Doug Bateman and Ken James made various measurements of the Great Clock. The team accurately measured the driving weights and pendulum and calculated the weights of each from the specific gravity of cast iron, which they had experimentally obtained from a broken fragment of the clock. Their results were:

Item	lb	kg
Going weight	560 (5 cwt)	254
Hour-striking weight	2376 (1 ton)	1180
Quarter-striking weight	2376 (1 ton)	1180
Pendulum bob	506 (4½ cwt)	230
Pendulum complete	707 (6¼ cwt)	322

FIG 17.39 The double-barrelled crab.

A report in *The Engineer* of October 1856 stated that the pendulum complete weighed 682 lb. Denison's first drawing of a three-legged gravity escapement says that the pendulum was 686 lb. Both are very close to the figure calculated by Bateman and James, which excluded the suspension spring and chops.

Bateman and James also measured the Q factor of the pendulum. The Q factor is a number that can be measured in an oscillating system to give the quality of the oscillator; the higher the number, the better the oscillator. The Q factor is most commonly used in electronics but is a useful measure of a precision clock. Doug and Ken measured the Q factor to be 9,300. As a comparison, the Q factor of a balance wheel on a watch is about 300 and that of a quartz crystal around 500,000. At 9,300, the Q factor of the Great Clock is similar to pendulums on precision regulators. The results were written up in the *Horological Journal* of February 1977.

CHAPTER EIGHTEEN
ICONIC BIG BEN

Big Ben's iconic image has been used in almost every aspect of music, literature, art, merchandising, and so on. To give a complete survey of its use would far exceed the space available, so just a few examples will be given.

MUSIC

Not surprisingly, the chimes of Big Ben feature in music. The little-known composer Ernst Toch (1887–1964) fled his native Germany in 1933 when he perceived the probable path of the political situation. Between 1933 and 1934 he stayed in London. Shortly after his visit he wrote *Big Ben: Variation-Fantasy on the Westminster Chimes for Symphony Orchestra, Opus 62.* Although little known, this piece has an exceptional charm.

Toch wrote of his work,

Once on a foggy night whilst I was crossing the Westminster Bridge, the familiar chimes struck the full hour. The theme lingered in my mind for a long while and evolved into other forms, always somehow connected with the original one. It led my imagination through the vicissitudes of life, through joy, humour and sorrow, through conviviality and solitude, through the serenity of forest and grove, the din of rustic dance, and the calm of worship at a shrine; through all these images the intricate summons of the quarterly fragments meandered in some way, some disguise, some integration; until after a last radiant rise of the full hour, the dear theme like the real chimes themselves that accompanied my lonely walk, vanished into the fog from which it had emerged.

A whole range of popular music and songs from 1900 to World War II used the Big Ben image. Known examples are: *Big Ben is Saying Goodnight* by Alan Murray; *When Big Ben Chimes* by Russell and Taylor; *The Big Ben Chime Waltz* by Pola, Hilton, and Steininger; and *Big Ben*

FIG 18.1 *The Westminster Chimes.* Music for handbells, dedicated to Sir Edmund Beckett Denison.

Song by Pontet and Clough. No doubt there are many others. One special piece of music is a polka, *The Westminster Chimes*, by the Royal Handbell Ringers. The work was dedicated to Sir Edmund Beckett, so must date between 1874 and 1886. One wonders what Sir Edmund thought about the illustrated cover that showed, as well as the Clock Tower, a miniature of the first Big Ben by Warners being tested.

Perhaps the most famous of its time, though largely forgotten today, was A.P. Herbert's light opera *Big Ben*, which appeared in 1946. Today we would call this a musical. The story is about a well-off owner of a department store: his son stands for Parliament in one constituency and a shop worker stands in the neighbouring constituency. Of course, they are on opposing political sides, and, of course, both are elected; they fall in love and marry, but not before being locked up in the Clock Tower for obstructing the House of Commons. The background is, of course, post-war, both in politics and the social situation, and as such the work is no longer topical. However, time smooths all things, perhaps it will reappear as a musical one day.

PHILATELY

Since the introduction of the Penny Black stamp in 1840, the British postal system stuck firmly to traditional designs for stamps, showing the head of the monarch and the postal value. By 1924, stamps were issued to commemorate the British Empire Exhibition at Wembley. There were stamps for the Silver Jubilee in 1935 and, shortly after, the coronation in 1937. The first appearance of the clock tower was a Cinderella for the 1937 Coronation. A Cinderella is a commemorative stamp

that has no postal value and is used on envelopes for decoration.

Strangely enough, Big Ben, the tower, the clock movement or the bell never featured on stamps on their own, not even for the centenary. Conferences used images of the Houses of Parliament and a special Big Ben cancellation was used for the 1957 Parliamentary Conference. However, the Royal Mail finally became aware of the market for stamp collectors and made amends by issuing a special commemorative sheet of ten different Big Ben stamps to celebrate the 150th anniversary. Oddly enough, Mongolia issued a stamp for the 1990 Stamp Exhibition long before the Royal Mail featured the Clock Tower.

MERCHANDISING

So many items have been branded with Big Ben or depicted its icon. Scarves, teapots, boot polish, beer, ties, T-shirts, and bands are just a few.

POSTCARDS

Thousands of different postcards featuring Big Ben must have been published. Rotary Postcards published a series of cards depicting unusual occupations in London. One card captured the Big Ben telescope man as he sits under the statue of Boadicea. It seems that he charged a fee for tourists to use his telescope to inspect the details of the Clock Tower. His telescope points up at the clock dials and it appears to have been an astronomical model but fitted with an eyepiece to turn the image up the right way. A pair of chained binoculars hangs from the telescope stand. For sale are nougat and chocolate bars, postcards, and maps. The man sits on a stout wooden box;

FIG **18.2** The first appearance of the Clock Tower in stamp form.

FIG **18.3** *London Life's Big Ben Telescope Man.* A postcard from a series published about 1910.

perhaps he used this to pack his equipment and wares so they could be taken home.

HP SAUCE

Perhaps the most famous product to feature Big Ben is HP Sauce. This was created by Frederick Gibson Garton, who was a grocer in Nottingham. He brewed the sauce to his own recipe and registered the name HP Sauce in 1896 after hearing that a restaurant in the Houses of Parliament had begun serving his product. In a shrewd move, he linked the sauce to the Houses of Parliament (HP) and used the Clock Tower image from 1908 to confirm the association.

CHAPTER NINETEEN
THE 150TH ANNIVERSARY CELEBRATIONS

THE RUN-UP TO THE 150TH ANNIVERSARY

Replica Bells

On 10 April 2008, the Whitechapel Bell Foundry cast a limited edition of 75 miniature Big Bens to commemorate the casting of the Great Bell 150 years previously. The casting took place on the same spot of the foundry and at the same time that the original bell was cast. Each bell was polished and inscribed with the name of the buyer. Some were given away at presentations, the others sold to collectors. News media were there in abundance to witness the casting, the *Ringing World* of 2 May 2008 recorded the event.

The British Horological Institute's 150th Anniversary

In June 2008, The British Horological Institute celebrated its 150th anniversary. Edmund Beckett Denison, designer of the Great Clock and Big Ben, was the Institute's President for 38 years. It was fitting that, at the Summer Show, the Palace of Westminster Clockmakers hosted a stand displaying various artefacts. Broken parts and

FIG 19.1 A miniature Big Ben cast on 10 April 2008, exactly 150 years after the original bell.

original wooden patterns were displayed, as well as a large number of pictures. The show was very busy and the Great Clock stand attracted a lot of interest.

FIG 19.2 The Westminster stand at the British Horological Institution Summer Show.

FIG 19.3 Left to right: in the background, Huw, Paul, and Ian, Clockmakers at the Palace of Westminster. In the front, three visitors to the show: Callum and Lewis are holding the bronze spanner used to remove the octagonal retaining nut on the minute hand, while Tyler holds the winding handle employed to wind the going train. This figure is reproduced in colour in the colour plate section.

150 YEARS

As a celebration of 150 years of service, Big Ben was treated to a good deal of publicity in 2009. A special 150th logo was launched and all visitors to the Clock Tower were given a special 150th lapel badge, showing the logo. The parliamentary website (www.parliament.uk/bigben) presented excellent information on the Clock's history, including the animations used in the visitor room. The website also included photographs and historical images, details of events during anniversary year, downloadable ringtones, desktop wallpapers, screensavers, and e-cards, as well as podcasts giving access to inside the tower.

A computer game 'Race Against Chime' was created with the Parliamentary Education Service as a fun way for youngsters to learn about Big Ben. The player has to control a man hanging on a rope with the objective of cleaning the dials with a sponge. As players progress through the levels, the wind becomes more of a problem and there is more dirt on the dials. Entry to each level is gained on answering a question about Big Ben's history, but it does not matter if players get it wrong; they get in all the same. Rumour has it that on completion of all levels, the player is treated to a massive firework display.

Numerous foreign newspapers and media caught the excitement of the event and they too produced their celebration of the birthday. Even the Google 'doodle' for the day displayed Big Ben.

More Media Coverage

- Broadcast media coverage during the year included:

- *BBC Breakfast*: New Year's 'leap second',
- *Blue Peter*: Time change weekend,
- *The Paul O'Grady Show*,
- BBC News: a day of coverage on 10 July, including items on Breakfast News, 1 p.m., 6 p.m. and coverage on the BBC News Channel (News 24) and London Tonight (ITV),
- Radio coverage: BBC Radio 4 presented a 15-minute musical—a specially commissioned piece of music commemorating Big Ben,
- Extensive press coverage around key anniversaries (31 May, 11 July)—in *The Times, The Sunday Times*, the *Telegraph*, the *Metro*, the *Daily Mail*, the *Daily Express*, the *Mirror*, and the *Evening Standard*,
- Magazines: specialist journals, such as the *Horological Journal*, plus a range of other titles, including *BBC History Magazine* and a feature in *The World of Interiors*,
- Foreign press and broadcast coverage, including Russia, China, the USA, Australia, Brazil, Germany, Spain, France, Netherlands, and Finland,
- Extensive coverage in regional press and radio broadcasts,
- Extensive online coverage.

THE GUY FOX HISTORY PROJECT

Big Ben is used to full protection and strict rules govern who is allowed to visit the old man. Thanks are due to the Guy Fox History Project, who gained entry for a whole year-4 class (8- and 9-year-olds) from St George's

Primary School in Camberwell. This visit inspired the children to create some unique drawings, included in a wonderful little 36-page booklet *Happy Birthday, Big Ben!* It is free, which is a nice change from the expensive horological books we now see being published. The booklet is very well produced and combines a wonderful mix of the children's drawings, correct historical information, and old engravings.

The Guy Fox History Project is a charity that seeks to engage children with the past to prepare them for the future. It produces project material about London and local history, which is intended to show the culturally diverse heritage of the City; it also seeks to emphasize the contributions made by individuals of various backgrounds to the arts, government, and the life of the City. Further, they want to encourage children to think about their own place in history and to consider what their contribution will be. The Project also offers workshops, which combine art, heritage, and information technology to develop educational resources that are distributed to other children.

The 'Happy 150th Birthday, Big Ben' Project was launched on Tuesday, 19 May 2009. The clockmakers and other staff from the Palace of Westminster were at London's City Hall along with Guy Fox staff, helpers, volunteers, teachers, and, of course, the children. It was nice to see other supporters there, such as the printer of the booklet and a firm of sponsoring solicitors. Cameras clicked when a very large and very furry Guy Fox appeared; no doubt the person inside was soon very hot with all the action and shaking children's hands as they were presented with their certificates.

Apart from *Happy Birthday, Big Ben*! Guy Fox published a Big Ben fact sheet, a sheet entitled *How to Make a Bell*, a DVD, a super facsimile of *The Illustrated London News* giving a compendium of *Illustrated London News* articles on the clock, and a teacher's pack of resource material. The DVD contains two mini documentaries, showing visits to the Clock Tower and to the Whitechapel Bell Foundry. Both are engagingly narrated by children. As a whole, it's a good example of how history can be made to involve and capture the imagination of youngsters. Support for the project came from Heritage Lottery funding.

FIG 19.4 The Guy Fox history book.

To be fair, it was decided that George Airy, the Astronomer Royal, would write a specification.

He invited 3 clockmakers to submit their plans and then decided which clockmaker would make the Great Clock.

> The clock should be accurate to within ONE SECOND of time. It will be the largest and most accurate public clock in the world!

George Biddell Airy

Edmund Beckett Denison

Edmund Denison helped George Airy decide.

Of the designs, the one by Edward Dent was the best. He and his stepson Frederick were chosen to build the clock. Edmund Denison suggested changes to improve their design.

> Edmund Denison invented a new type of escapement which made the clock very accurate.

Edward Dent

Frederick Dent

The Dents started building, with Denison's help.

The Illustrated London News, 7th February 1857. Courtesy of the London Metropolitan Archives.

FIG 19.5 An example of the booklet's engaging style.

MR SPEAKER'S RECEPTION

On Tuesday, 2 June 2009, Mr Speaker hosted a reception in the State Rooms at Speaker's House; the reception was held to celebrate the 150th anniversary of the Great Clock 'Big Ben'. Tim Treffry, the President of the British Horological Institute was there, along with BHI Members Ian Westworth, Paul Roberson, and Huw Smith, the team of clockmakers who look after the Great Clock, and Mike McCann, the Keeper of the Great Clock. There were about 80 guests, including Clock Tower staff past and present, horologists, and campanologists, Members of Parliament, House of Commons staff, and external dignitaries and guests, who had contributed to anniversary celebrations. The reception was held in the State Dining Room, with all its rich and intense gold and red Gothic-revival decoration, the work of Augustus Pugin.

Mr Speaker said a few words of introduction and mentioned that living next to the clock tower had never kept him awake at nights. He then welcomed Edward Garnier, QC, MP, who is a direct descendant of one of Lord Grimthorpe's sisters. Edward introduced Teddy Beckett, the fifth and present Lord Grimthorpe; he reported that they were both the great-great-great-great nephews of Edmund Beckett Denison, the first Lord Grimthorpe, who designed the clock and the bell. Teddy Grimthorpe is descended from E.B. Denison's brother. He outlined the family involvement with Parliament, which was not insignificant and spanned many generations, and proceeded to give a short but succinct speech on Edmund Beckett, later Lord Grimthorpe, full of family history.

Speaking about Edmund, the designer of the Great Clock and bells and past President of the BHI for many years, Edward said:

He was unquestionably a difficult and a controversial man but, I think, a brilliant and determined man. A gifted mathematician, a highly successful QC, an engineer, locksmith, clockmaker and horologist, a church builder and restorer and amateur architect, a businessman, landowner and member of the House of Lords; and a prolific writer on clocks, architecture and abstruse religious subjects. His first published work was entitled 'Six letters on Dr Todd's Discourse on the Prophecies Relating to Antichrist in the Apocalypse'. It must have been a real page-turner.

He also was known for writing a vast number of letters to The Times. To quote his biographer, Peter Ferriday, 'He had an unequalled knack of bringing out the worst in his fellow men. His own letters to the press were brutish, bullying and libellous… He believed in speaking out, and if anyone disagreed with him the only inference to be drawn was that man was either an idiot, an habitual liar, or had been bribed.'

He applied this robust attitude to his dealings with Sir Charles Barry and anyone else who had or thought they had a proper interest in the clock tower and its contents. One recipient of his letters wrote, 'The tone of Mr Denison's last letter relieves me from the task of continuing to notice his remarks.'

Despite his reputation for rudeness and belief in his own genius, both of which were to a large part justified, he has, with others, left us with a truly magnificent clock, which still after 150 years, and thanks to the dedication and care of those who have looked after and continue to look after this great but delicate machine, keeps time to within one second's accuracy, and a bell, Big Ben, that booms out hourly above us and around the world. This was a man who can truly be described as a polymath, a Victorian renaissance man but he also someone who believed he possessed the Englishman's right to do what he likes. We live in a different time but thanks to him we can measure its advance second by second.

After Edward finished his speech, the Speaker, Edward Garnier, and Lord Grimthorpe unveiled a plaque to commemorate the 150th anniversary of the Great Clock. This will be installed in the belfry.

Six young competition winners were at the reception; this was a national competition, in which children had been tasked to design a Christmas card on the theme of 'Parliament and Big Ben' to celebrate the anniversary. MPs were invited to involve schools in their constituencies in the competition and over 400 entries were received. The national winners of the competition attended the Speaker's reception. Regional winners received certificates and badges to congratulate them on their achievements. Mr Speaker presented the national winners' prizes at the reception and their winning cards were on display. The cards were a delightful mix of cartoon and reality and exhibited budding talent. The national winners' other prizes included a tour of the Clock Tower, meeting the Keeper of the Great Clock and the clockmakers, and a 'Winners' Tea' with their MPs.

With the official business over, a considerable amount of social networking amongst old friends and new acquaintances followed, all facilitated by champagne and canapés. As the night took over from the day, and with the bells ringing above as they have for a century and a half, the guests agreed that it had been a most memorable event.

COMMEMORATIVE MERCHANDISE

As with any important event, a selection of merchandise supports the celebration. The Palace of Westminster launched a variety of items: a mug, fridge magnet, notebook, postcards, and pencils, all showing the 150th logo.

In 1959, a question for the Postmaster General was asked in the House of Commons; whether a Big Ben anniversary stamp was to be issued. The answer was No. However, for the 150th anniversary, the Royal Mail produced a set known to stamp collectors as 'smilers'. A smiler is a stamp where a special design is joined to a normal postage stamp: the stamp with the special design has no postal value. The Big Ben Smiler issues were only available as a single sheet of ten first-class stamps. Designs showed a detail of the clock movement, the Ayrton light, a view from the tower, Pugin, the dial, an old engraving, the 1834 fire, the design for the palace, the hour bell, and top of the Clock Tower. The Royal Mail also brought out an anniversary medal cover; set into the card cover was a medal showing the bell Big Ben. The Big Ben stamp is cancelled with a Big Ben frank mark.

A £5 coin was struck by the Royal Mint. Although the coin is a commemorative issue for the 150th anniversary of Big Ben, it also doubles up as merchandising for the 2012 London Olympics. The obverse shows a perspective view of the dial and has a blue Olympics logo; this is the first time that colour printing has been used on a British coin. The coin bears the enigmatic slogan *Nations touch at their summits*.

150TH EXHIBITION AT THE BRITISH HOROLOGICAL INSTITUTE

Following the success of the Westminster stand at the British Horological Institute's 150th show, the team returned in June 2009 to stage another display. This time an 8-ft high model of the clock tower, complete with working dials and chimes, was displayed. A computer showed animations of the escapement and striking, and young visitors had a go on the 'Race Against Chime' game. Once more, the ever-popular display of wooden patterns and broken parts generated a lot of visitor interest.

FIG **19.6** An original wooden pattern used to make castings of the bevel gears that connect the clock to the dials.

FIG **19.7** The wooden pattern from which the pendulum bob was cast.

BIG BEN RESTORED

Big Ben, that is the real Big Ben, the bell, had a facelift. Those who recall its aged grey-green patina might just bring to mind the flecks of paint it carried, the by-product of several generations of steelwork painting. Added to that, there was the odd spot of guano left by the occasional avian visitor who had managed to slip into the bell tower, evading the tight security system.

A conservation project was planned that would not only improve the bells' appearance but, more importantly, would protect them from the weather and any associated deterioration. The conservation clean also meant that the bells were thoroughly inspected.

Prior to deciding on the method of conservation, the Estates Archivist carried out some research and found that pre-World War II the bells were cleaned and polished with blacklead once a fortnight. This cleaning stopped during the war and the bells' appearance deteriorated. Immediately after the war, they appear to have been cleaned and coated with preservative oil. However, regular cleaning did not appear to have been re-established after war and by the centenary in 1959 the bells were described as dirty and spattered with pigeon droppings and paint. The bell foundry that cast the bells was also consulted and confirmed that the traditional finish applied to bells is a wax-based graphite-black grate polish.

The process began with a full inspection of the bells and some advance testing by the specialist conservator. The inspection revealed that the bells were in good condition, with an overall green patina, which is a result of natural weathering. However, the bells did have a number of very old paint drips and blemishes, which needed removing. The conservator also suggested that a protective surface treatment against further weathering should be applied.

A number of different wax treatments and finishes were tested and the Conservation Architect, curator, and clock team, together with the conservator, agreed that the wax-based graphite polish was the most acceptable finish. The work was carried out at the end of May by a team of three trained conservators and was monitored by the Conservation Architect, curator, and clock team. All surface dirt and oil was removed, along with paint drips and guano. This was followed by an application of wax-based graphite polish that was then buffed up to a shine. The work did not interrupt the operation of the clock or the chiming of the bells, but scaffolding had to be erected to provide access to the bells. At the end of the work, the bells looked truly magnificent.

This is not the first time that Big Ben has had a bit of a spruce up. When it was first installed in 1858, the Clerk of Works had the bell bronzed with nitric acid to make it look nice. When the crack in the bell was discovered in October 1859, Edmund Beckett Denison accused the Whitechapel bell foundry of concealing defects by applying a coloured wash. He was proved wrong and had to retract his comments.

ENTHUSIASTS' WEEKEND

Over the weekend of 11–12 July 2009, an 'Enthusiasts' Weekend' was held to celebrate Big Ben's 150th anniversary. Four hundred keen clock and bell-ringing fans, successful in a ballot for tickets via specialist journals such as the *Horological Journal*, were welcomed to Parliament for tours of the Clock Tower and a glimpse behind the scenes.

The Palace Clockmakers and the Clock Tower Tours Team worked hard over a hectic weekend,

FIG **19.8** The 150th lapel badge given to all visitors to the Clock Tower in 2009.

FIG **19.9** The date plate on the Clock.

running eight tours per day, ensuring that all visitors enjoyed their special visit. Westminster Hall's Jubilee Café was opened so that visitors could have a well-deserved cup of tea after climbing the 334 steps to the top of the tower.

Feedback from the event was very positive, with visitors praising the weekend's smooth organization, the knowledge, and expertise of the tour guides, and the friendliness of the staff. One visitor commented particularly that the whole thing 'ran like clockwork'.

On the evening of 11 July, the message 'Happy Birthday Big Ben, 150 years, 1859–2009' was projected onto the Clock Tower after sunset.

PORTCULLIS HOUSE EXHIBITION

From September to November, an exhibition entitled 'Parliament's Clock: Big Ben at 150' was staged in Portcullis House. Works of art from the Parliamentary Collections and from public and private lenders were displayed to tell Big Ben's story and to explain its role as symbol of the nation. The exhibition included portraits of Benjamin Hall and Edmund Denison, a drawing of the planned Houses of Parliament, a fine lithograph of the first Big Ben, an ink drawing of the Great Clock, and the silver trowel used to lay the Clock Tower's first stone in 1843.

London's Open House Weekend on 19 September attracted thousands of visitors to Portcullis House, who were able to visit the Clockmakers' display stall, as well as the exhibition. On 20 September, the Open House Weekend activities transferred to Westminster Hall. On both days, an actor in full Victorian garb, portraying Sir Benjamin Hall, MP, was on hand to engage with visitors and interpret the Big Ben story.

OTHER EVENTS

Other events were staged around the country and abroad. The Big Ben model at Legoland in Windsor was emblazoned with a birthday banner on 31 May and 11 July. A chocolate Clock Tower was enjoyed by the British Chamber of Commerce in Hungary, an ice model was displayed in Covent Garden, and a 70-ft high clock tower made of 500 bales of straw was erected near the A51 in Cheshire.

CHAPTER TWENTY
CONCLUSION

THE CLOCK

The design of the Great Clock, with its unique escapement, produced significant changes in the British horological world. From 1860 onwards, the new flatbed design was used almost exclusively for all medium and large turret clocks, and Denison's double three-legged gravity escapement was the norm on good-quality clocks. Edmund Beckett Denison continued as President of the British Horological Institute until his death in 1905. Even a year or so before his death, Edmund was still corresponding with churches and the like, freely giving advice on clocks and, sometimes, on bells.

There was a great social and economic boom in the period from 1850 to 1900 and new buildings were put up to satisfy the need for factories, hospitals, prisons, town halls, and the like. As towns and cities spread, the railway provided a new and fast connection. In the 1880s, Great Britain adopted a standard time based on Greenwich, in response to a great need to make sure that railway timetables worked. The whole growth process was complex and interlinked. One small strand of the development was the bringing of accurate timekeeping to the community, ensuring that meetings and communication worked better and more efficiently.

THE BELLS

Big Ben is still one of the largest bells in the country and quite the most famous. Bell founding is a very specialized art, which even today employs materials much the same as were used in mediaeval times. It is doubtful that casting the Great Bell made any significant contribution to founding technology, but what it did achieve was to boost the popularity of the Westminster (originally, Cambridge) quarters. Bell founding prospered as town halls, churches, etc., put in clocks that sounded the quarters. Often a ring of eight or more bells would be supplied to churches, all part of the boom years.

THE TOWER

Probably the biggest contribution the Clock Tower made was to popularize the illuminated clock dial. With gas being more readily available, new buildings could have a dial lit up at night. The free-standing clock tower in the square of a market town or city suburb became commonplace. Clock tower architecture was very varied but some designers adopted the Westminster tower as their starting place. At London's Victoria station, a clock tower called 'Little Ben' was installed in 1894; it stands about 30 ft tall, including the wind vane.

Just how the image of the Clock Tower became the icon to represent the Nation is not known; it was a gradual process. As printing developed, more people came to know Big Ben. The booming postcard industry in the early 1900s did much to popularize the tower, and newspapers, too, became more graphic. When movies emerged, people would see the tower or hear the chime to get the message, '*The scene is now in London.*'

Apart from being the Nation's icon, Big Ben also represents Westminster with all its meaning of history, Parliament, democracy, and free speech.

Long Live Big Ben!

FIG 20.1 Big Ben's Clock Tower has become a symbol for London, and the nation.

APPENDIX A
WEIGHTS AND MEASURES

When Big Ben was built, metric measurements were not used in Britain. Here is a summary of the units that would have been used, together with their metric equivalents.

WEIGHT

Unit	Contained	Metric equivalent
One ton (t)	20 hundredweight (cwt)	1016 kg
One ton (t)	2240 pounds (lb)	1016 kg
One hundredweight (cwt)	4 quarters (qr)	50.80 kg
One hundredweight (cwt)	112 pounds (lb)	50.80 kg
One quarter (qr)	28 pounds (lb)	12.70 kg
One pound (lb)	16 ounces (oz)	0.453 kg
One ounce		0.0283 kg

DISTANCE

Unit	Contained	Metric equivalent
One mile (m)	1,760 yards (yd)	1.609 km
One yard (yd)	3 feet (ft)	91.44 cm
One foot (ft))	12 inches (in)	30.48 cm
One inch (in)		2.54 cm

CURRENCY

Unit	Contained	Metric equivalent
One pound (£)	20 shillings	£1.00
One guinea (gn.)	21 shillings	£1.05
One shilling (s.)	12 pennies	£0.05
One penny (d.)	2 half pennies (½ d.)	£0.004 (approx.)
One penny	4 farthings (¼ d.)	£0.004 (approx.)

APPENDIX B
CLOCK TOWER DATA

Item	Number	Imperial	Metric
Tower			
Height to top		316 feet	96 m
Height to dial centre		180 feet	55 m
Width		40 feet	12 m
Steps to Clock Room	290		
Steps to Belfry	334		
Steps to lantern (the Ayrton Light)	393		
Amount of stone used		30,000 cubic feet	850 cubic metres
Amount of bricks used		9,200 cubic feet	2600 cubic metres
Number of floors	11		
Locations of building materials	Anston, Yorkshire Caen, Normandy, France Clipsham, Rutland (for restoration work in 1983–5)		
Clock dials			
Number of clock dials	4		
Diameter of clock dial		22½ feet	7 m
Length of hour figures		2 feet	0.6 m
Clock dial frames	Cast iron		
Number of pieces of glass in each clock dial	312		
Dial illumination	28 energy efficient bulbs at 85 W each		
Minute hands			
Material	Copper sheet		25 kg, including counterbalance
Weight		55 lb	
Length		14 feet	4.2 m

Distance travelled by tip of minute hands per year		118 miles	190 km

Hour hands

Material	Gunmetal		
Weight		165 lb	75 kg each, including counterweights
Length		8¾ feet	2.7 m

Bells

Big Ben

Weight		13½ tons	13.7 tonnes
Height		7.2 feet	2.2 m
Diameter		8.8 feet	2.7 m
Musical note	E		
Hammer weight		440 lb	200 kg

First quarter bell

Weight		1 ton	1.1 tonnes
Diameter		3.6 feet	1.1 m
Musical note	G#		

Second quarter bell

Weight		1¼ tons	1.3 tonnes
Diameter		3.9 feet	1.2 m
Musical note	F#		

Third quarter bell

Weight		1¾ tons	1.7 tonnes
Diameter		4.6 feet	1.4 m
Musical note	E		

Fourth quarter bell

Weight		3.9 tons	4 tonnes
Diameter		5.9 feet	1.8 m
Musical note	B		

Clock

Length of frame		15 ft 6 ins	4.7 m
Width of frame		4 ft 11 ins	1.5 m
Estimated weight		5 tons	5.1 tonnes
Going train weight		5 cwt	254 kg
Hour-striking train weight		1 ton	1180 kg
Quarter-striking train weight		1 ton	1180 kg
Pendulum bob weight		506 lb (4½ cwt)	230 kg
Pendulum complete		686 lb (6¼ cwt)	311 kg

KEY SOURCE MATERIAL

The following principal sources were used to prepare much of the content in this book:

Denison, Edmund Beckett (later Sir Edmund Beckett and then Lord Grimthorpe) (1850–1903). *A Rudimentary Treatise on Clocks, Watches, and Bells*, (edns 1 to 8).

Mercer, Vaudrey (1977). *Edward John Dent and his Successors*. Antiquarian Horological Society.

Mercer, Vaudrey (1983). *A Supplement to Edward John Dent and his Successors*. Antiquarian Horological Society.

Includes five pages on Big Ben; mentions the cause of the 1976 disaster. This has four folding plates in the back showing Denison's sketches for the Great Clock.

James, Jabez (1859). On the Process of Raising and Hanging the Bells in the Clock Tower, at the New Palace, Westminster. *Minutes of the Proceedings of the Institution of Civil Engineers*, **19** (1860), 3–20.

Antiquarian Horology.

Horological Journal.

The Illustrated London News.

The Times.

Hansard.

Letters concerning the clock and bells in the National Archives and Parliamentary Archives. Many were published in various Parliamentary Returns listed below:

Return of 13 July 1847 for Copies of the Papers and Correspondence Relating to the Great Clock and all other clocks in the New Palace at Westminster (Mr Edmund Denison).

Return of 2 March 1848 for Copies of the Papers and Correspondence Relating to the Great Clock and all other clocks in the New Palace at Westminster (Mr Wyld).

Return of 7 June 1855 for Copies of the Papers and Correspondence Relating to the Great Clock and Bells for the New Palace at Westminster (Mr Wyld).

Return to the Hon House of Commons 24 May 1852. Copy of the Memorial presented to Her Majesty's Commissioner of Works and Public Buildings by the Clockmakers' Company of London respecting the Great Clock to be erected at the New Palace at Westminster; together with the answer thereto (Mr Octavius Morgan).

Return to the Hon. House of Commons 11 June 1852. Copies of the Papers and Correspondence Relating to the Great Clock for the New Palace at Westminster.

A Portion of the Papers relating to the Great Clock for the New Palace of Westminster. Printed by Order of The House of Lords with Remarks, 1848. For Private Circulation Only (Vulliamy's comments added to the House of Lords Return).

BIBLIOGRAPHY

The following is a list of books relating to the Great Clock and to turret clocks in general. A brief comment has been added to help students assess the useful of the book's contents since some of the books are long out of print and difficult to find.

Airy, George Biddell (ed. by Airy, Wilfrid, BA, MInstCE) (1896). *Autobiography of Sir George Biddell Airy, KCB, MA, LLD, DCL, FRS, FRAS, Honorary Fellow of Trinity College, Cambridge, Astronomer Royal from 1836 to 1881.* Project Gutenberg ebook 10655.

Students of Victorian science will find this book fascinating: it is a mixture of Airy's diaries with additional private letters added by his son Wilfrid. These give a good picture of what Airy was doing plus an insight into his private life. Horologists will find reference to the Observatory clocks interesting.

Becket, Edmund: *see under* Lord Grimthorpe.

Beeson, C.F.C. (1971). *English Church Clocks 1280–1850.* Antiquarian Horological Society [Also published by Brant Wright Associates, 1977].

The only modern book dedicated to turret clocks. An absolute must for the enthusiast, even though it stops at 1850.

Brayley, E.W. and Britton, J. (1836). *The History of the Ancient Palace and the late Houses of Parliament at Westminster.* John Weale, London.

Good pre-fire history and engravings.

Cocks, Barnett (1977). *Mid-Victorian Masterpiece. The Story of an Institution Unable to Put its own House in Order.* Book Club Associates.

A history of the Palace of Westminster.

Collis, Len (1986). *Rupert and the Trouble with Big Ben.* Dragon Books. [Also published by Carnival, 1989].

A book for children, where Rupert Bear and Bill Badger are asked by Big Ben for a 2 p. piece to help it keep time. Astonishingly accurate drawings of the movement. Ideal for 5-year olds and as an introduction to turret clocks!

Dakers, Andrew (c. 1941). *The Big Ben Minute.* Andrew Dakers, Ltd.

Darwin, John (1986). *The Triumphs of Big Ben.* Robert Hale.

First-hand information on the disaster of 1976 and the reconstruction.

De Carle, Donald (1959). *British Time.* Crosby Lockwood, London.

A chapter on Big Ben. Just right for those who don't want to go too deep.

Denison, Edmund Beckett: *see under* Lord Grimthorpe.

Dent, Frederick (1855). *Treatise on Clock and Watch Work, with an Appendix on the Dipleidoscope.* Edinburgh.

A reprint of the article with this title from the eighth edition of *Encyclopaedia Britannica*, written by Edmund Beckett Denison.

Dodd, George (1843). Description of a visit to the factory of Messrs Moore & Co. In *Days at The Factories*, pp. 305–25.

Recounts a visit to a church clock factory.

The Evening News (1959). *Big Ben, 1859–1959*. Birthday souvenir. 36 pp.

Ferriday, Peter (1957). *Lord Grimthorpe 1816–1905*. John Murray, London.

A must for any Big Ben fan, or for those to whom Grimthorpe's crusty temperament appeals.

Gillgrass, Alfred (1946). *The Book of Big Ben*. Herbert Joseph, London.

A useful history of the clock, just right for the newcomer.

Gillgrass, Alfred (c. 1942). *Big Ben—The Signature Tune of the British Empire*. Leeds.

Gordon, G.F.C. (1925). *Clockmaking Past and Present. With Which is Incorporated the More Important Portions of 'Clocks Watches and Bells', by the late Lord Grimthorpe, Relating to Turret Clocks and Gravity Escapements*. 1st edn, Crosby Lockwood & Sons, London [Trade edn, 1928; 2nd enlarged edn, Technical Press Ltd, London, 1949].

Extracts on turret clocks from Denison's book.

Grimthorpe, Lord. Edmund Beckett Denison was born 1816. In 1874, he inherited a Baronetcy and became known as Sir Edmund Beckett. In 1886, he was raised to the Peerage and became Lord Grimthorpe.

Grimthorpe's Rudimentary Treatise. This book was undoubtedly the most popular book that Grimthorpe wrote. It went through eight editions: Grimthorpe went through three

different names, and the title of his work was likewise changed several times. Since this has often been a source of confusion, here are all the editions listed in full with title and current name of the author.

Denison, Edmund Beckett, MA (1850). *A Rudimentary Treatise on Clock and Watch-making: With a Chapter on Church Clocks, and an Account of the Proceedings Respecting the Great Westminster Clock*. John Weale London. 279 pp. [Price 2/–].

Denison, Edmund Beckett, MA, QC (1855). *Clock and Watch Work*. From the eighth edition of *Encyclopaedia Britannica*, London. 144 pp.

In his introduction to the fourth edition, Denison explains that this is really the second edition of *A Rudimentary Treatise*.

Denison, Edmund Beckett, MA, QC (1857). *Clock and Locks. From the* Encyclopaedia Britannica, 2nd edn. Adam & Charles Black, Edinburgh.

Denison says in his preface, '...may be regarded as a third edition on the Rudimentary Treatise'.

Denison, Edmund Beckett, MA, QC (1860). *A Rudimentary Treatise on Clocks and Watches, and Bells; with a Full Account of the Westminster Clock and Bells*, 4th edn, John Weal, London. 434 pp. [Price 3/6].

In his introduction, Denison says that he wrote the article in the eighth edition of *Encyclopaedia Britannica*. Exactly the same book also appeared signed 'Frederick Dent'.

Denison, Edmund Beckett, MA, QC (1868). *A Rudimentary Treatise on Clocks and Watches, and Bells; with a Full Account of the Westminster Clock and Bells*, 5th edn. 424 pp. [Price 3/6].

Reprint of the fourth edition with minor updates. In the front is an advertisement from the publisher explaining how the 5th edition was reprinted in 1867, without updates, and without Grimthorpe's authority. The publishers explain how this happened and apologize for the action. The new edition had minor updates and additions; the new appendix was made available as a separate book for the convenience of those who already had the fourth edition.

Denison, Edmund Beckett, LLD, QC, FRAS (1868). *An Appendix to the Fourth and Fifth Editions of* A Rudimentary Treatise on Clocks and Watches and Bells. Virtue & Co., London, pp. 379–417 [Price 1/–].

Beckett, Sir Edmund, Bart, LLD, QC, FRAS (1874). *A Rudimentary Treatise on Clocks and Watches, and Bells*, 6th edn. Crosby Lockwood & Co., London. 384 pp. [Price 4/6].

Beckett, Sir Edmund, Bart, LLD, QC, FRAS (1883). *A Rudimentary Treatise on Clocks and Watches, and Bells*, 7th edn. Crosby Lockwood & Co. London. 400 pp. [Price 4/6].

Grimthorpe, Edmund Beckett, Lord, LLD, KC, FRAS (1903). *A Rudimentary Treatise on Clocks and Watches, and Bells for Public Purposes*, 8th edn. Crosby Lockwood & Co. London. 404 pp. [Price 4/6].

Grimthorpe, Edmund Beckett, Lord, LLD, KC, FRAS (1974). *A Rudimentary Treatise on Clocks and Watches, and Bells for Public Purposes*. EP Publishing. 404 pp. [Facsimile reprint of the 1903 8th edn].

Denison, Edmund Beckett, MA (1848). On Clock Escapements. *Proceedings of the Cambridge Philosophical Society*. Read 27 November.

Denison, Edmund Beckett, MA (1849). On Clock Escapements. *Transactions of the Cambridge Philosophical Society*, **VIII**, 639–41.

Denison, Edmund Beckett. MA (1849). On Turret Clock Remontoires. *Transactions of the Cambridge Philosophical Society*. **VIII**, 663–8.

Denison, Edmund Beckett, MA (1853). On Some Recent Improvements in Clock Escapements. *Proceedings of the Cambridge Philosophical Society*. Read 7 February.

Denison, Edmund Beckett (1856). *Lectures on Church Buildings*, 2nd edn.

A total of nine pages on turret clocks, mostly as they relate to buildings. Conducts a tirade in true Grimthorpian style, as may be expected, against the traditionalism of clockmakers and the lack of attention paid by architects to clocks.

Beckett, Sir Edmund (1865). *Astronomy Without Mathematics*, 1st edn. [3rd edn, 1867: 6th edn, 1876].

Nothing horological in this.

Beckett, Sir Edmund, Bart (1876). *A Book on Building, Civil and Ecclesiastical*, 1st edn. 362 pp.

Hill, Rosemary (2007). *God's Architect. Pugin and the Building of Romantic Britain*. Allen Lane.

The Illustrated Exhibitor.

Published in 1851 as a serialized catalogue of the various exhibits at the Great Exhibition of 1851. No 23 contains the engraving of Dent's clock, designed by Denison.

The Institution of Mechanical Engineers (1981). *Big Ben—Its Engineering Past and Future.* Institution of Mechanical Engineers.

Some history not often publicized, plus a scientific look at the cause of the failure that caused the 1976 disaster.

Jones, Christopher (1983). *The Great Palace. The Story of Parliament.* BBC.

Macdonald, Peter (2004). *Big Ben, The Bell, the Clock and the Tower.* Sutton Publishing.

Useful for recent history of the clock.

Phillips, Alan (1959). *The Story of Big Ben.* HMSO.

Pickford, Christopher J. (ed.) (1995–2009) *Turret Clocks: Lists of Clocks from Makers' Catalogues and Publicity Materials.* AHS.

Rawlings, A.L. (1948). *The Science of Clocks and Watches,* 2nd edn. Pitman Publishing. [3rd edn, BHI, 1993].

Good on horological theory from a scientific approach. Mentions Grimthorpe's gravity escapement.

Ross, John (1987). *Big Ben and the Clock Tower.* HMSO.

Schoof, W.G. (1898). *Improvements in Clocks and Marine Chronometers.* London.

Starmer, William Wooding. *Quarter-Chimes and Chime Tunes.* Novello and Company.

Vulliamy, Benjamin Lewis (1828). *Some Considerations on the Subject of Public Clocks, Particularly Church Clocks: With Hints for their Improvement.* London.

Vulliamy, Benjamin Lewis (1830). *A Supplement to the Paper Entitled* Some Considerations on the Subject of Public Clocks, Particularly Church Clocks: With Hints for their Improvement *Consisting of a Correspondence with the Committee for Building the New Church at Bermondsey, on the Subject of a Clock for that Church; with Preliminary Observations, and Additional Papers.* London.

Vulliamy, Benjamin Lewis (1844). *A Supplement to the Paper Entitled* Some Considerations on the Subject of Public Clocks, Particularly Church Clocks: With Hints for their Improvement *Consisting of a Correspondence with the Committee for Building the New Church at Bermondsey, on the Subject of a Clock for that Church; with Preliminary Observations, and Additional Papers. To Which is Added Messrs Donkin and Bramah's Report, on the Clock at New Church Bermondsey, made by Messrs Moore and Son, Dated April 18 1829.* London.

Vulliamy, B.L. (1831). *A Brief Summary of the Advantages Attendant upon the New Mode of Construction of a Turret Clock, as Adopted by B.L. Vulliamy FRAS, Clock-Maker to the King, to the Honorable Board of Ordnance, and to the New Post Office: or in Other Words, Reasons Why a Clock Made on this Construction, will Measure Time More Accurately, and is More Durable, and More Easily Kept in Order, and Consequently Less Expensive, (Without Reference to the Execution of the Work), than a Clock upon the Old Construction.* London.

Vulliamy, Benjamin Lewis (1837). *Testimonials respecting Clocks made by B.L. Vulliamy.* London.

Vulliamy, Benjamin Lewis (1846). *On the Construction and Theory of the Dead Beat Escapement.* John Oliver, London.

Vulliamy, Benjamin Lewis. *A Portion of the Papers Relating to the Great Clock for the New Palace at Westminster.* Printed by order of the House of Lords, with remarks, for private circulation.

Vulliamy, Benjamin Lewis (1851). *A Statement of the Circumstances Connected with the Removal of B.L. Vulliamy by the Commissioners of Woods from the Care of Such of the Government Clocks as are in their Custody; Together with Copies of all the Correspondence Connected with the Same.* London.

White, W. Douglas (1958). *The Whitehurst Family. Derbyshire Clockmakers Before 1850.* Derby.

INDEX

Entries in *italic* refer to illustrations
Entries in ***bold italic*** refer to illustrations that also appear in the colour plate section

Printed and bound by CPI Group (UK) Ltd, Croydon, CR0 4YY